T0390676

DRAGONFLIES AND DAMSELFLIES OF THE WORLD

A GUIDE TO THEIR DIVERSITY

Klaas-Douwe B. Dijkstra

PRINCETON UNIVERSITY PRESS
PRINCETON AND OXFORD

This book is dedicated to the popularization of Odonata, especially to Dennis Paulson and Bert Orr, who contributed more to that cause in their part of the world than anyone else (p. 251). I also wrote it in memory of Jill Silsby, who inspired us all with the publication of her landmark *Dragonflies of the World* almost a quarter century ago.

Published in 2025 by Princeton University Press
41 William Street, Princeton, New Jersey 08540
99 Banbury Road, Oxford OX2 6JX
press.princeton.edu

Conceived, designed, and produced by
The Bright Press
an imprint of The Quarto Group
1 Triptych Place, London, SE1 9SH, United Kingdom
T (0) 20 7700 6700
www.quarto.com

Library of Congress Control Number 2024941708
ISBN: 978-0-691-25503-3
Ebook ISBN: 978-0-691-25504-0
British Library Cataloging-in-Publication Data is available

Publisher **James Evans**
Editorial Director **Isheeta Mustafi**
Art Director and Cover Design **James Lawrence**
Senior Editor **Joanna Bentley**
Project Editor **Katie Crous**
Design **Wayne Blades**
Picture Research **Katie Greenwood**
Illustrations **John Woodcock**

Cover photos: Front cover, clockwise from top right: Shutterstock/guentermanaus, Flickr/Budak, Charles J. Sharp, Flickr/Budak, Cynthia Su, Graham Winterflood, David Smith, Keith Wilson. Back cover: Charles J. Sharp.

Printed in Malaysia

10 9 8 7 6 5 4 3 2 1

CONTENTS

INTRODUCTION

ABOVE | Unlike for bees or beetles, English has no non-technical and unambiguous term for members of the order Odonata, whose name translates as "toothed ones." The word "dragonfly" refers to any odonate (including damselflies!) to some people and just to the suborder Anisoptera to others (p. 14), of which a female Common Thorntail (*Ceratogomphus pictus*) in South Africa is shown here. This book strictly uses "dragonfly" in the latter sense and "odonate" for members of the order as a whole.

Spirits dwell among us. For months, even years, they lurk underwater, but on warmer days they emerge into our lives on land. Borne on lacelike wings, these small creatures may still be easily missed. The watchful eye, however, will soon see their bright colors shoot by everywhere.

Worldwide, thousands of varieties exist, each bound to its own watery realm. From the darkest pool to the clearest spring, the widest lakes and the greatest rivers. From the farthest island to your backyard.

If we would celebrate their needs and diversity as we do our own, we would care for every river. As all land drains into a river, and every river into the sea, we would care better for the land and sea as well and, therefore, for all life. The world would be habitable for every being and for everyone.

You might know these spirits as dragonflies and damselflies. Just as we can only share our knowledge and beliefs as words and stories, life expresses itself as individuals and species. Each is packed

with meaning, responding and conforming to an everchanging world so we might understand it.

The only challenge is to translate this form and variety to our own language. The transliteration of life has barely begun, however. Of possibly 10 million species of animal, plant, and fungus, at most a fifth has been named. Generally not in one of the 7,000 living languages, though, but in Latin. And that is just a name!

Encouragingly, ever more people see dragons and damsels for their value. It is a cultural revolution, really: for all of history, we generally did not even consider most of these species worth their own words, yet in mere decades field guides were written and common names coined for many of them. Now thousands of people share their findings every day, each one refining our understanding of every species further.

Together, over 6,400 species known are placed in the order Odonata, called odonates colloquially, or more poetically, odes. This book introduces the language they provide that allows us to listen to the landscape. How did their diversity, and the enchantment and insight they offer, arise from the change that has gone before? How, moreover, do odonatists the world over translate that language into their own?

ABOVE | Unlike "dragonfly" (see caption left), the word "damselfly" is unambiguous, referring only to members of the odonate suborder Zygoptera, as exemplified by the male Short-tipped Bluestreak (*Lestoidea brevicauda*) from Australia.

FRESHWATER FLIERS: THE ORIGIN OF ODONATES AND THEIR DIVERSITY

When rain falls, or we dig a pond, dragonflies and damselflies soon appear. As water keeps flowing or wetland forms, further species establish. When sources are fouled, banks are stripped bare, or weather patterns shift, some species shy away. But, as conditions recover, the early responders rapidly return.

What brings these species there and how do they survive? How did their abilities and preferences arise and shape the diversity we see? And how did the species' varied appearance, which allow us to appreciate their diversity so easily, come about?

We do not have answers for all of over 6,400 known odonate species, but the taxonomic profiles should give an idea of the forces at play. First, however, we must address what is special about odonates as lifeforms and about freshwater as a living environment. While that requires describing the basics of their behavior and ecology, Dennis Paulson's *Dragonflies & Damselflies: A Natural History* is recommended for further details.

LIVING FOSSILS? WHAT MAKES ODONATES SPECIAL

Odonates are often presented as ancient, almost primordial, insects. While many extant species may only have arisen in the last few million years, their general habits and form are certainly very conserved evolutionarily. All are thought to derive from a strictly terrestrial air-breathing creature whose immature stage adapted to life underwater, perhaps 400 million years ago (mya). While over 50 insect lineages invaded freshwater from land, this is the earliest (surviving) one.

Almost every descendant of that lineage returned to land and air only as adults. Like virtually all insects, they never went back to the sea from which the insect ancestor emerged nearly half a billion years ago. Unlike many insects, however, they never attained complex relationships with other species groups—for example, as pollinators or parasitoids—nor even with each other, such as by caring for offspring or making nests.

Of extant insects, only mayflies may have arisen from the same freshwater invasion, but they then diverged so long ago (over 360 million years) that actually proving that is hard. While not a single mayfly adult feeds, their nymphs' diets range from detritus to unicellular algae and (more rarely) other insects. Odonates are obligate carnivores in both stages of life, by contrast, barely even specializing in their mostly invertebrate prey.

LIFE CYCLE OF AN ODONATE

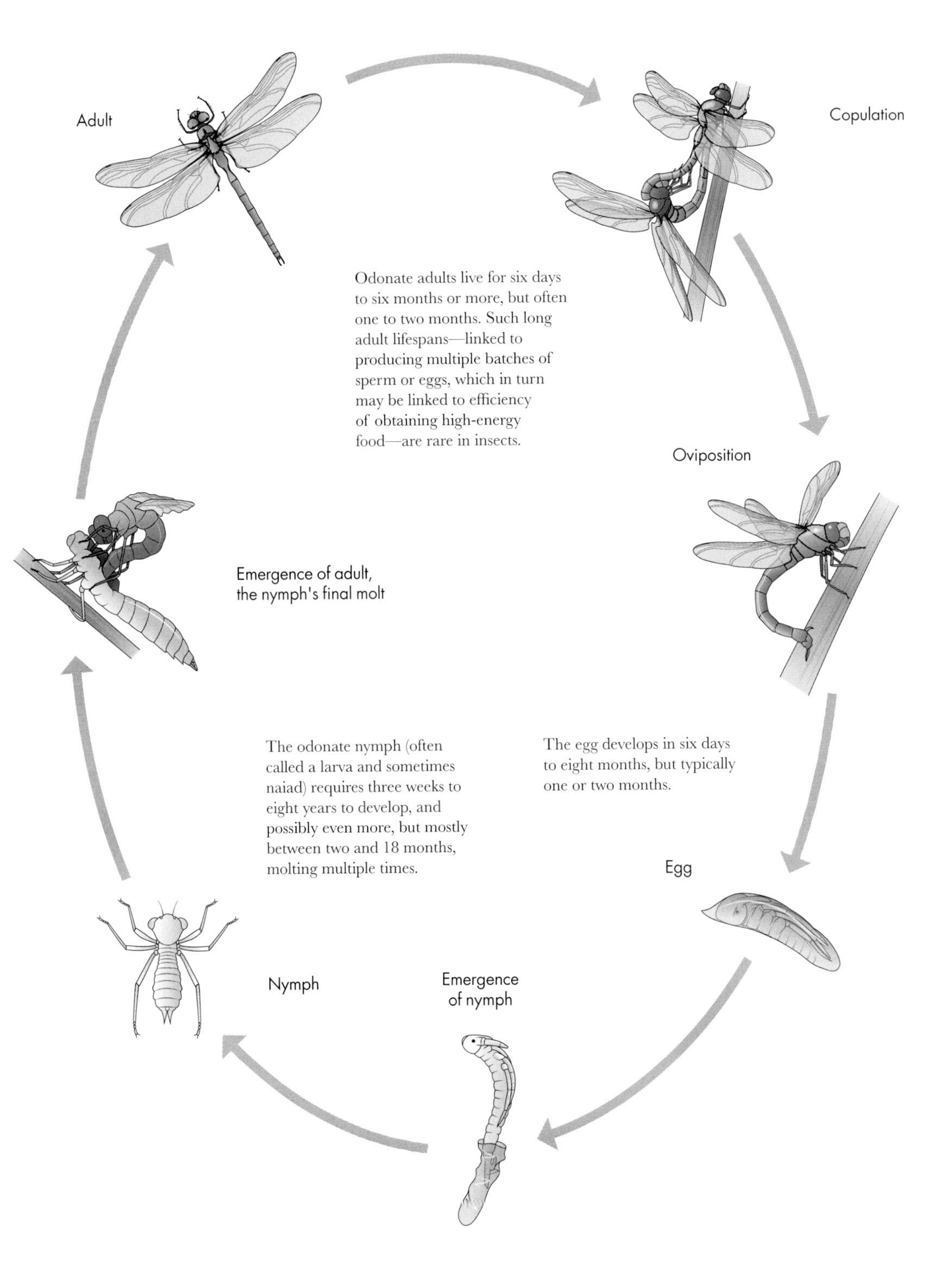

Why did habits change so little? The answer must lie in the group's unique features. The mouthparts of all odonate nymphs—but no other lifeform—are modified into a grasping arm, which can strike in 0.005 seconds, thrice as fast as a mantis's forelegs can. Such specific equipment may not only be better for hunting underwater than any other, but might also not remodify readily for feeding in other ways.

Similarly, the adults' flight and vision may imply both inescapable constraints and an eternal head start for an aerial hunter. Independent stroking of the four wings, for example, allows for high acceleration, speed (up to 22 mph/36 kmph or 10 in/25 cm per beat) and maneuverability, and thus for diverse flight modes and hunting strategies (see p. 12 for more on vision). Without a pupal stage in which to radically reconfigure the gut, odonates may also be committed to a similar diet as both nymph and adult.

ABOVE | Being double-hinged and partly elastic, the nymphs' labium (lower lip) can be slung forward like a catapult. The tip of this "mask" bears two graspers, called palps, with fingerlike hooks and often teeth. Most groups (such as this damselfly nymph) have a flat mask, used rather like a chameleon's tongue, that of many dragonflies is scooped, more like a pelican's beak (p. 86).

WHY WATER IS LIFE

With at least 100,000 species, a tenth of Earth's animal diversity is believed to live in freshwater. Despite their constrained ecological repertoire, around 5 percent of these are odonates. These habitats, though, cover less than 1 percent of the planet. How can there be so many species associated with them?

Many basic requirements for life, of course, are already there in abundance. Water is life's main solvent and, with carbon dioxide and sunlight, the key ingredient for photosynthesis. As liquid, water also moderates temperature fluctuations and concentrates resources such as nutrients, intensifying the productivity within it even further. As it flows and settles, moreover, water carves and sculpts an endless array of microhabitats into the landscape.

In odonates, that potential is expressed in the nymphs' morphological diversity. The individual's success at attaining adulthood, after all, is determined largely by its ability to feed and avoid being fed on within its underwater confines, often for considerable time. A wild stream with see-through water and plenty of oxygen, but just rocks to hide under, for example, demands a very different anatomy than a murky pond with various living and dead vegetation to lurk among.

THE SHAPE OF WATER: BOUNTIFUL ISLANDS IN A SEA OF LAND

As it flows to the lowest point, water on land is highly fragmented. Effectively, each waterbody is like an island overflowing with food and possibility, locked in by land. Either because they cannot leave or do not need to—flow keeps most streams and rivers replenished, for example—many organisms there easily become isolated and diverge genetically, potentially leading to new species. Branching like a family tree, a river system can literally be a genealogy.

ABOVE | Among European dragonfly nymphs (from top left to bottom right), the cylindrical River Clubtail (*Stylurus flavipes*) digs into sand, while the flatter and spiderlike Splendid Cruiser (*Macromia splendens*) slides under leaf-litter. The Lilypad Whiteface (*Leucorrhinia caudalis*) suspends itself among pond vegetation, but the Ringed Cascader (*Zygonyx torridus*) hangs on in rapid water. Among damselflies, the gangly Beautiful Demoiselle (*Calopteryx virgo*) lives among rootlets under the riverbank, while the Odalisque (*Epallage fatime*) hides beneath stream rocks.

Such a high-energy and opportunity-rich yet isolation-prone environment, in other words, is primed for life to diversify. This explains how the mere 29 percent of Earth not covered by sea can harbor 51 percent of all fish species, despite being largely hostile to them. Similarly, odonates are richest in tropical forests, where abundant water and high temperatures limit primary productivity least, particularly in varied terrain such as mountains, where localized species will almost always find reliable water and their preferred niche nearby (see pp. 41 and 233 for examples).

LIVING LIKE WATER: LIFE MUST LEAVE FRESHWATER TO FLOURISH THERE

While freshwaters are exceptionally amenable to life, they are challenging habitats to survive in too, depending on an intense dynamic of evaporation, precipitation, and flow to persist, while also being especially sensitive to changes on Earth's surface and in the atmosphere. Fortunately, water not only fragments by flowing to the lowest point, but also creates a huge interface with the land, air, and sea.

Appropriately, the lifecycles of most freshwater species (including all odonates) closely mirror the water cycle, rising like vapor from the surface as adults, but always returning there to breed (p. 7). This allows them to access resources above water too, to evade adversity, and even to take advantage of opportunities that arise with constant change.

For example, odonates can profit from warm water, abundant prey, and few competitors in drying pools to grow fast as nymphs because they can escape these confines as adults, flying as far—or staying out of water as long—as is necessary to find suitable habitat again (e.g. p. 238). Their populations may not diverge as easily as those of odonates in running waters (see above) but have better chances of avoiding extinction.

FRESHWATER: PRECIOUS AND CONNECTED

The concentration of biodiversity in inland waters, such as South Africa's Palmiet River, is hardly surprising: to the three-quarters of species on Earth that live on land, including all human beings, freshwater is the most precious substance, and here that most limiting resource is limited least. These waters and most of their lifeforms, furthermore, bind all habitats together: many insects (including odonates!) emerge onto land and into the air, of course, while many fish are born in a river but grow up in the sea. The sad corollary is that freshwaters are also the most threatened living environment (p. 30).

RIGHT | This female Ochre Spreadwing (*Lestes ochraceus*) in Madagascar must dwell inconspicuously in dry grass for long periods, only being able to breed once the monsoon starts filling up puddles.

As habitat is scattered in time and space, species can even persist just by chance. Two very similar species can avoid outcompeting each other as long as both find an opening somewhere. Who finds a spot first, moreover, influences the levels of competition, predation, and resources the next arriving species encounters. They set the scene for the next arrival. Not only is every species at least slightly different, so is every opportunity.

STAY OR GO? CHANGING HABITATS LEADS TO DIVERSITY

The rates by which species disperse, arise, and survive vary strongly with their habitat. These can balance out, though: weak dispersers will stumble onto distant opportunities more rarely, but more species may evolve when they do (p. 19). Numbers in both standing and flowing waters are indeed high, with about 65 percent of species preferring the latter.

Almost all extant species strongly favor one or the other and yet odonates switched between these habitats many times during their evolution: most stream odonates on remote islands, for example, evolved from pond-dwellers arriving from afar (pp. 136–141). This suggests that when such rare shifts do occur, these contribute substantially to diversity. Perhaps due to an advantage acquired from the faster-paced life in ponds, for example, most still-water groups include some running-water specialists on the mainland too (p. 150). Inversely, several running-water groups broke into still waters, possibly because they lacked such large predators, opening a whole new world of opportunity to them (p. 103).

ABOVE LEFT | Expanding in drier periods, deserts threaten all freshwater life. Sands deposited across Africa in the past, however, retain and slowly release seasonal rainfall today, providing stable yet nutrient-poor habitats for specialists such as Gabon's Redwater Leaftipper (*Malgassophlebia andzaba*).

ABOVE | Odonate eyes are among the largest in insects, composed of up to thousands of ommatidia, each with its own lens and optic nerve. Odonates have more different light-sensitive proteins than we do, moreover. These must allow them to see well underwater as nymphs, but especially to perceive the adults' exceptional diversity of colors.

Such events, often coming in the wake of great environmental change, probably contributed disproportionately to the exceptional success of insects such as odonates. Mastering powered flight before pterosaurs, birds, and bats, while also being smaller, their ability to access opportunity and avoid danger and difficulty was never equaled. They thus not only became multicellular life's most varied offshoot, but established Earth's highest species density in freshwaters: up to 80 percent of animal species there are believed to be insects.

RIGHT | Melanin and other pigments in the adults' living cells scatter light, thus creating their brightest but also most evanescent colors, such as the red of Africa's Superb Dropwing (*Trithemis hartwigi*). Melanin in the cuticle is more durable, appearing matt to glossy black or even strongly metallic, as on its face. Reflective waxy scales excreted epidermally (called pruinosity or pruinescence) give the red its violet hue, but can also appear white to bright blue and dark gray, depending on their density and background color.

FLYING COLORS: DIVERSITY THROUGH BEAUTY

Aside from its choice of habitat for the fertilized eggs, and thus offspring, individual reproductive success is governed by adult mate choice. Although females make their own choices (pp. 135–7)—and can be showy too (p. 90)—and while features such as claspers also matter (p. 99), the males' colorfulness draws most attention in that arena (pp. 35 and 180). Often emphasized by display behavior, much of this splendor is to impress not females, but rival males. There is no choice of sites and mates, after all, without first having gained access to them.

These sexual signals can reinforce genetic divergence attained by other means. Having developed different coloration in isolation, for example, individuals from two river systems may no longer recognize each other as potential mates (or rivals) when they meet, yet physically still be able to interbreed. Other factors, such as background color (pp. 55 and 163) or the presence of similar-looking species (p. 183), affect these signals' effectiveness too.

The nymphs of an African *Pseudagrion* species (photo p. 151) of lush ponds, for example, had to adapt to very different conditions on Lake Tanganyika's beach-like shores. The structurally identical adults can move freely between these habitats, however, countering any genetic divergence. The males differ markedly in coloration, though, probably affecting their visibility—to mates, rivals, and predators—in each landscape and providing the final push for their populations' definitive separation as species.

Ultimately, countless factors interact to determine what the odonates we see look like. Some of the largely separate forces shaping the nymphs (pp. 8–9) can carry over to the adults (pp. 94, 156, and 204). Inversely, male conflicts may favor ever broader wings and bigger bodies (p. 180), affecting adult feeding or dispersal behavior, but also the nymph's development time and thus survival chances. Together, these factors induced an overwhelming variety of colors and shapes, but also confusing similarities in unrelated groups (pp. 148, 180, and 196).

DYNAMIC DUO: THE RISE OF DRAGONFLIES AND DAMSELFLIES

By fathoming the character and interdependencies of dynamic duos such as freshwaters and insects, aquatic nymphs and aerial adults, and stream- and pond-dwellers, we may begin to understand how life can be so varied. How, even within the constraints of a flying freshwater carnivore, species can range from 1-month lifecycles crossing oceans (p. 52) to tunneling into the cold seeps of New Zealand's Alps for a decade (p. 110). Or how two similar ponds or streams, even within eyeshot of each other, may be inhabited by quite different sets of species.

Also in their classification, odonates suit our binary way of thinking (but see p. 130). Half the species are "true" dragonflies, with two pairs of different wings. That suborder's name, Anisoptera, is Greek for "unequal-wings." The others, with two identical pairs, are not "false" dragonflies, but damselflies or Zygoptera, "yoked-wings." Referring to balance, both "yoke" and "zygon" come from a prehistoric root for "join" or "unite," from which the word yoga stems as well.

TOP | The male Slender Jewel (*Stenocypha tenuis*), a damselfly limited to central Africa's forest streams.

LEFT | Basking male Harlequin Darner (*Gomphaeschna furcillata*), a dragonfly restricted to bog pools in North America.

BELOW | Griffinflies represent one of various extinct orders in the superorder Odonatoptera. Members of Meganisoptera (formerly known as Protodonata) were not just many times larger than today's dragons and damsels, with wingspans of almost 30 in (70 cm), but also lacked most key characteristics (p. 16). Fossils nonetheless suggest their nymphs already possessed the unique prehensile mask. As some were the largest insects to ever live, their extinction is often linked to lowering ambient oxygen levels. Most griffinflies were not gigantic, though, so competition with modern odonates may have accelerated their demise too.

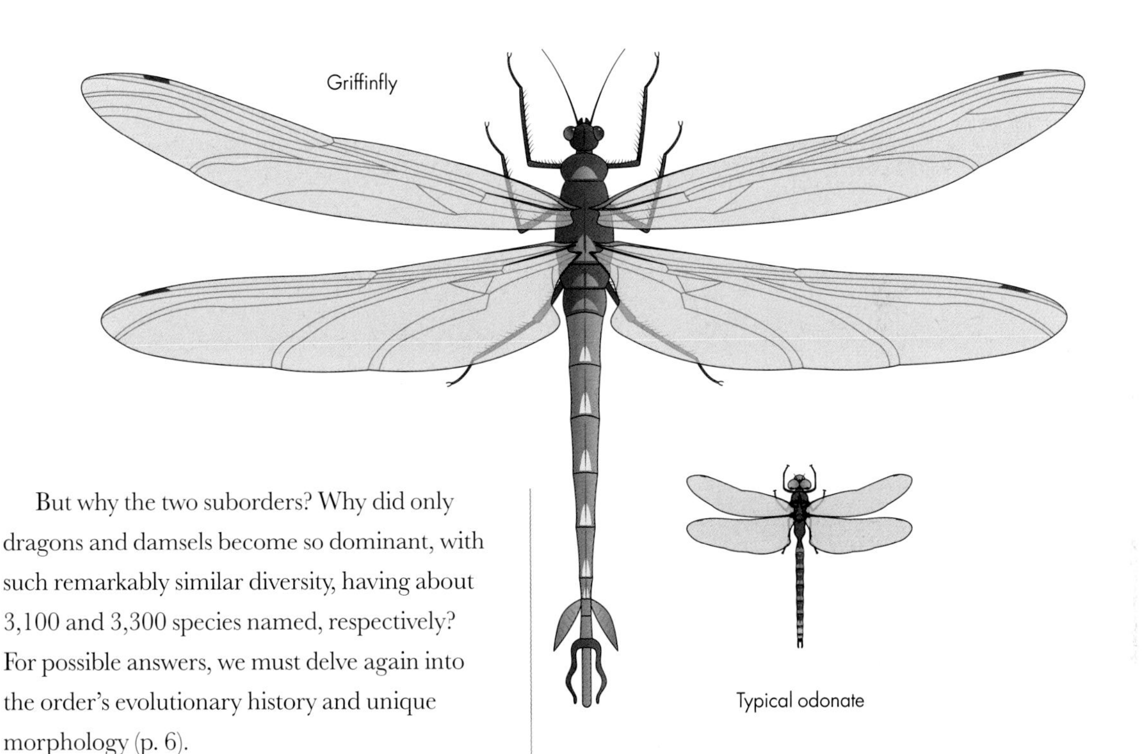

But why the two suborders? Why did only dragons and damsels become so dominant, with such remarkably similar diversity, having about 3,100 and 3,300 species named, respectively? For possible answers, we must delve again into the order's evolutionary history and unique morphology (p. 6).

GOING THEIR SEPARATE WAYS: ORIGIN OF THE SUBORDERS

Thinking of ancient dragonflies, we often picture bird-sized monsters flying in forests of giant horsetails and lycopods. Although none of these behemoths survived beyond the Permian, which ended with the greatest extinction crisis ever, the first dragonfly-like insects with a contemporary look did arise during their reign, at least 280 mya.

Dragons and damsels diverged in the Triassic, some 50 million years later, roughly when the first mammals arose (placental mammals only appeared 90 million years later, though). Diversifying contemporaneously with the dinosaurs, most of the extant odonate families probably existed by the end of the Cretaceous.

As numerous fossils and a few surviving so-called Anisozygoptera attest (p. 131), various groups with mixed characters once existed. A fossil from the Jurassic suggests that the distinct respiratory modes of dragon and damsel nymphs (next section) could even be combined. The dominance of just the two groups, therefore, may only have come about gradually, over tens of millions of years.

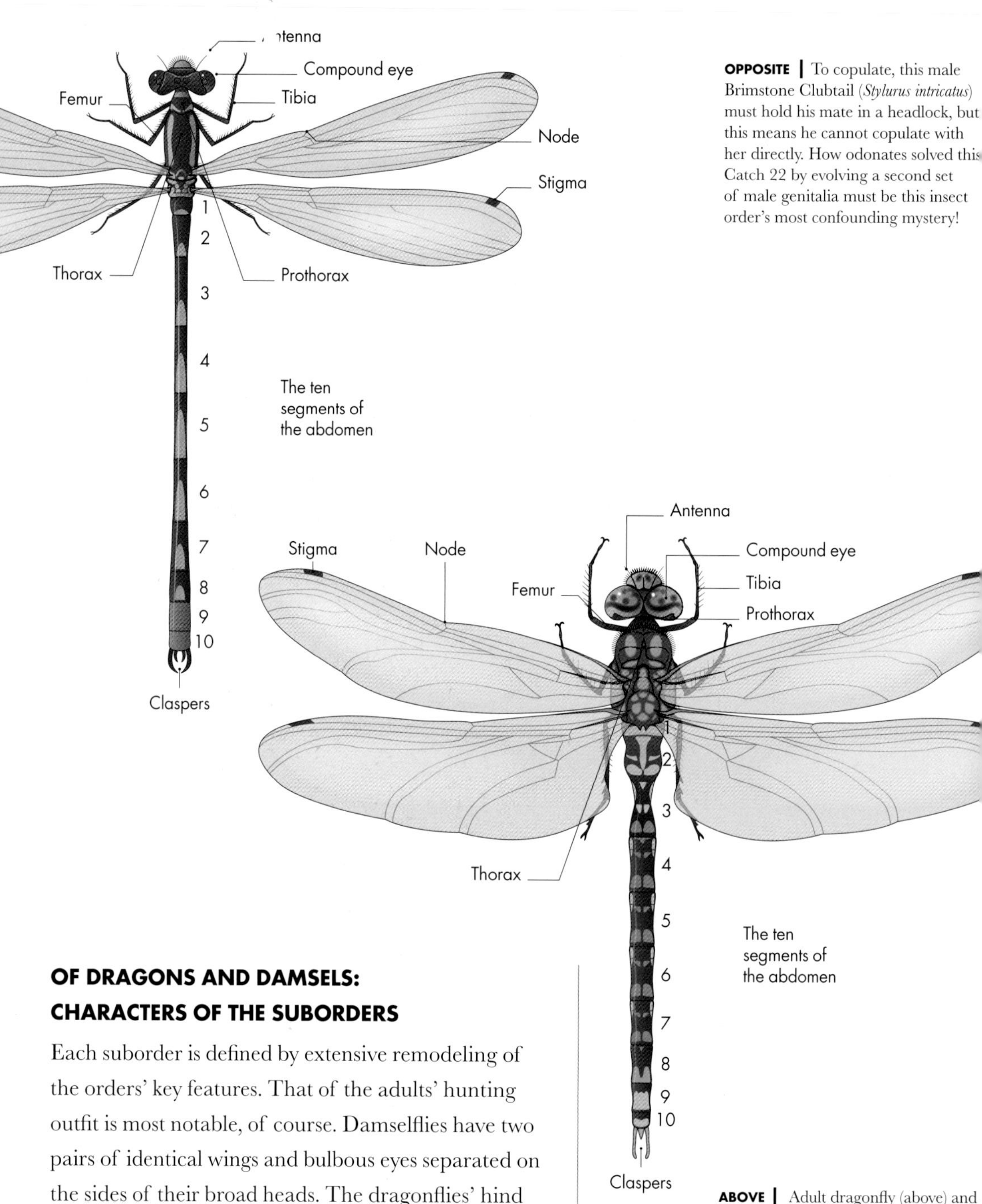

OPPOSITE | To copulate, this male Brimstone Clubtail (*Stylurus intricatus*) must hold his mate in a headlock, but this means he cannot copulate with her directly. How odonates solved this Catch 22 by evolving a second set of male genitalia must be this insect order's most confounding mystery!

ABOVE | Adult dragonfly (above) and damselfly (top left) males, showing the differently placed eyes and the former's broader hindwings. Unlike precursors such as griffinflies (p. 15), modern odonates have unsegmented appendages on the abdomen tip (forming claspers in these males) and small antennae. They also have nodes, fortified breaks to provide flexibility in the wing's leading edge like an airplane's flaps, and stigmas serving as stabilizing counterweights.

OF DRAGONS AND DAMSELS: CHARACTERS OF THE SUBORDERS

Each suborder is defined by extensive remodeling of the orders' key features. That of the adults' hunting outfit is most notable, of course. Damselflies have two pairs of identical wings and bulbous eyes separated on the sides of their broad heads. The dragonflies' hind wings are considerably expanded, by contrast, especially their base, while the huge eyes envelop the head (compare p. 130).

The reconfiguration of the abdomen's unsegmented terminal extremities is even more significant. There are two pairs of appendages, the cerci above and paraprocts below, with the single epiproct in-between. All nymphs

have unremarkable cerci, but while the three other structures form leaflike gills in most damsels, which may also function as a tailfin (pp. 133 and 239), they form a shutter in dragons (p. 9). This anal pyramid closes off the rectum, which has pleated walls that serve as gills instead. Inhaled water can be expelled with force, so the gut can provide jet propulsion too!

The extremities are smallest in adult females, probably mostly serving a tactile function. In males, however, they form claspers to grasp their mates. The cerci act like two fingers, but while dragonfly males apply the epiproct as a thumb to grab females between the eyes, damselflies instead use the paraprocts like double thumbs to lodge onto the prothorax, the necklike front of the thorax directly behind the head (p. 18).

This headlock is the basis for odonates' unique reproductive behavior (p. 7). How the sexes meet varies widely, but copulation cannot commence until the male has clasped the female, forming a tandem. As both have their genital opening near the abdomen tip, and the male's tip is behind the female's head, they cannot connect directly, however!

All odonate males (but no other lifeform) therefore transfer their sperm to a small sack below the abdomen base (usually just) before copulation. Part of the so-called secondary genitalia, this sperm vesicle is long and segmented in dragons and thus also functions as a penis. In damsels, the ligula, a slide-like supportive structure, serves this purpose instead. Both carry structures to remove rival

sperm, however, as eggs are only fertilized when laid. Females may expel unwanted sperm themselves too.

To ensure they father her offspring, males therefore often accompany females after copulation and during oviposition. This also reduces the chance that the female (and thus his potential offspring) is preyed upon. As all damselflies have an ovipositor to painstakingly place eggs inside suitable substrates (see also p. 110), many species have contact-guarding, whereby the pair remain in tandem so that the male can pull the female away from danger or to another egg-laying spot. This organ is lost in over 80 percent of dragonfly species (pp. 86 and 93), however, so oviposition behavior is much more varied; males often perch or hover nearby, or do not guard at all.

LIVING APART TOGETHER: HOW TWO SUBORDERS MAY COEXIST

Perhaps only two odonate blueprints (dragonflies and damselflies) were ultimately successful because, with such specialized features, few combinations are actually optimal. But why two and not one?

Damsels are generally smaller and slighter, occupying tighter spaces than dragons. Most flights are short and low, and adults rest closer

TOP LEFT | While male dragonflies grab females by the head (p. 17), damselflies such as Common Bluets (*Enallagma cyathigerum*) connect at the female's prothorax.

LEFT | The female Blue Emperor (*Anax imperator*) in Europe lays her eggs alone, flying off at the slightest disturbance.

OPPOSITE | The female Orange-faced Sprite (*Pseudagrion rubriceps*) from India has descended underwater and is being guarded by her mate.

to their perches, often deeper in vegetation. This favors eyes that project past the perch and neatly folding wings, the latter also making them less evident to predators (p. 194). Nymphs must have smaller home ranges too, with little room for internal gills in their long and narrow bodies.

The dragonflies' roomier habits, by contrast, better suit the nymph's getaway by jetpack, and the adults' expanded wings (spread for rapid takeoff) and spherical eyes. Such big eyes hinder a four-point grip around the prothorax, so they grab between them instead, with one "thumb" fitting better than two.

The simplest explanation for the suborders' similar diversity is that both managed to largely fill the ecological space available to a group of predatory aquatic insects. Most sites occupied by odonates, indeed, have both dragons and damsels present. This hypothesis implies that, while the groups' habits are similar enough to provide comparable evolutionary potential, they differ sufficiently for one group to not affect the other disproportionately through competition or predation.

That is possible if they indeed feed and breed at different spatial scales, limiting their interaction, while the effect of dispersal takes place at a similar scale. Damsels move less on average than dragons, but, as with stream- and pond-living groups—which also occur side-by-side without impacting each other much—that may balance out in their species numbers (p. 12). Proportionately, there are indeed more dragonfly species in ponds—as well as in colder and drier regions—but more damsels at streams and in the tropics (compare pp. 129 and 233).

PARALLEL UNIVERSES: COMPARING THE SUBORDERS TO UNDERSTAND THE ORDER

Contrasting the two suborders' evolution helps fathom the order's diversity. While the most eccentric dragons survived in cooler climes (p. 129), for example, the many isolated damsel groups are mostly tropical (pp. 194–248). No damselfly group attained the dominance in running waters that Gomphidae did (p. 92), moreover, perhaps because their nymphs' blueprint did not allow for such a wide range of digging lifestyles.

Remarkably, the second-largest group at standing waters in each suborder belongs to the lineage that is related most distantly to the lineage that claimed most of the extant diversity there: Aeshnidae (p. 112) and Libellulidae (p. 34) among dragonflies, and Lestidae (p. 238) and Coenagrionidae (p. 132) in damselflies.

Perhaps only such evolutionarily distinct groups differ sufficiently to simultaneously be successful in these abundant but dynamic habitats (pp. 10–11). This suggests that many other lineages could live there too, and probably once did. These, then, must have largely vanished as the others swept across the globe. Odonates may seem like living fossils (p. 6), barely having changed since they arose hundreds of millions of years ago. Many of the species we see, however, probably only emerged in the last couple of million years.

ABOVE | Dragons and damsels commonly coexist, such as this Lilypad Whiteface (*Leucorrhinia caudalis*) and Common Bluet (*Enallagma cyathigerum*) on a Dutch pond, but use the landscape differently due to their distinct size and build.

RIGHT, FROM TOP | The vigilant male Armourtail (*Armagomphus armiger*) overlooking an Australian stream, the male Selys's Giant Spreadwing (*Orolestes selysi*) resting among leaves in Taiwan, and the female Common Spreadwing (*Lestes sponsa*) in a Scottish bog show how differently dragons and damsels use the space in the landscape, depending on their build and sex.

THE CITIZEN'S INSECTS: CREATING THE ODONATIST'S LANGUAGE

Odonates may be called "the birder's insects," but birds are really just "insect vertebrates," rerunning an evolutionary script perfected millions of years—and millions of species!—before them by combining flight with a modest size in order to access opportunity and evade threats (p. 12). As a result, both are found everywhere in great diversity, so their enthusiasts overlap substantially.

Such interest rarely begins for any practical or economic reason—because odonates devour mosquitoes, for example. Quite the opposite, actually: we seem to like them for their distance from our daily woes. If biodiversity is like language (p. 5), these beings are like music or (perhaps more obviously) colorful art, to which we relate intuitively without immediately realizing the impact or significance.

When enjoying odonates for their beauty or the serenity of their haunts, after all, we implicitly recognize the value of a diverse and intact living environment. Sadly, these waters are treated more like planetary plumbing than as the key to sustaining life. Freshwater may be lifegiving, but we can favor our own needs until all other life has been squeezed out. That, indeed, is the greatest challenge of being human: to realize always that everything that can serve us has its own value too.

Representing the most critical and connecting resource and environment on land, yet being likable purely for themselves, these ubiquitous insects are especially suited to help us navigate that challenge. Both the scientific name *Libellula*, the oldest for a dragonfly (p. 38), and Zygoptera, meaning all damselflies (p. 14), refer to balance. Ultimately, calling odonates "the citizen's insects"

might convey their potential best. Not only for the sense of ecological responsibility they could nurture, but because ordinary people are increasingly valuable to their study and popularization.

THE BIRDER'S INSECTS? WHY WE ARE NOT THERE YET

Even though their value is unparalleled in the ecological worldview that is (cautiously) emerging, odonates are largely irrelevant in the prevailing view focused on economics. This means that, while just as accessible as birds or wildflowers, they lack such groups' history (and thus culture) of use. Around the world, for example, most people are lucky to even have words for "dragonfly" and "damselfly" in their language to distinguish the many species around them.

Nonetheless, most who encounter another lifeform do not ask first after its biology, but for its name. Its place, in other words, in the human record. Typically, language evolves organically, words being coined, adjusted, and selected gradually. Publishers and users of field guides are not so patient, however. Of some 4,000 names available in English for 3,000 of over 6,400 odonate species today, 85 percent were minted in the last 25 years: authors simply had to create them from scratch. Even some in this book are new.

Therefore, while the technology already exists that would allow anyone to photograph an odonate anywhere, identify it using image recognition, and get instant feedback on its environmental significance, much of the knowledge and language needed to implement that dream has yet to be refined. This chapter shows how the gap is being closed, how the remaining imperfections thwart our understanding, and, most importantly, what anyone with an interest can contribute.

ABOVE | Spain's Cazuma Pincertail (*Onychogomphus cazuma*) was discovered by enthusiasts in 2017, the first species named from southwestern Europe since the nineteenth century!

LEFT | Most odonate adults can be identified from a photograph, but odonatologists must still often collect samples for research, as many taxonomic issues have yet to be resolved.

DISCOVERING ODONATE SPECIES

While over four-fifths of the millions of insect species that must exist remain nameless, most odonates have at least been described scientifically. In 2008, nearly 5,700 species were known, with between 1,000 and 1,500 estimated that remain to be found. Today, over 6,400 are known, so the final tally may lie above 7,500 species. An astonishing 380 species, for example, were added from the American tropics in the twenty-first century.

While most discoveries are still made by researchers in rainforests, anyone can find new species anywhere by looking carefully enough, as Spanish citizen scientists showed with *Onychogomphus cazuma* in 2017. Similarly, while this book's author works as an independent expert, his two coworkers on a study describing 61 new species from across Africa in 2015, investigate odonates in their free time.

DEFINING ODONATE SPECIES

As they are boldest and have the bright colors, distinctly shaped claspers, and genitalia that separate the species most easily, study focuses on adult males. As these differences are linked to reproduction (p. 13) and presumably limit interbreeding, most odonatists implicitly apply a "biological species concept," defining species as groups of populations whose members can produce viable offspring in nature.

Ultimately, however, any name applied to an individual remains a hypothesis, tested with as much (different) evidence as possible, whether morphological, behavioral, ecological, geographic, or genetic. Also when you record a dragonfly under a certain name, you contribute to that species' common definition, especially when sharing your effort, such as on databases like iNaturalist or Observation.org. Weighing the available evidence, after all, you concluded that a particular species occurs at a particular location at a particular time. And each identification, whether correct or not, influences the community.

ABOVE | Presenting reproductive barriers even between close relatives, the coloration and genitalia of the male (at the front of this tandem) serve as our main guides to separate odonate species.

RIGHT | This book's author had the honor to present Attenborough's Pintail (*Acisoma attenboroughi*) from Madagascar to the broadcaster on his 90th birthday.

NAMING ODONATES SCIENTIFICALLY

While determining the validity of a species is ultimately a judgment call, that of their names can be governed by rules, set for scientific animal names by the International Code for Zoological Nomenclature. The species name *Pantala flavescens* is formed by appending the specific epithet *flavescens* to the genus name *Pantala*. Combinations must be unique, so if synonyms (multiple names that describe the same concept) or homonyms (different concepts that have the same name) arise, the oldest name has priority.

Species' distinct regional forms are sometimes named as subspecies. By separating the differently colored Beautiful Demoiselles in southwest Europe as *Calopteryx virgo meridionalis*, for example, those elsewhere automatically became *C. v. virgo*. This practice is relatively rare for Odonata, though (e.g. p. 153).

Aside from using this binomial (trinomial in the case of subspecies) structure, a new name must be published with a description of the species named (preferably in a peer-reviewed journal) to be valid. Only a single specimen designated to exemplify that species, ideally kept in a museum, definitively links the name to the concept behind it, though. Almost every so-called holotype for Odonata is an adult male.

Technically, names are just labels. They are created for posterity, however, and all language has emotional resonance, so they deserve some care and creativity. While a quarter of odonate species are named for people and an eighth after places, names that describe some inherent property convey most meaning, such as *auripennis* for a golden-winged species.

Any name using the Latin alphabet can be valid, but most odonate genera are based on Greek and most specific epithets on Latin. These ancient languages are unfamiliar to many, so names such as *Denticulobasis* and *Priscagrion* put Latin words on a Greek base, while an epithet like *wenshanensis* is an English transliteration of a Chinese placename with a Latin suffix.

CLASSIFYING ODONATES

By appending its specific epithet to a genus name, the concept of a species is hung on to a nested structure of taxonomic ranks, like a child's name on a family tree. Aside from suborders (p. 14), the order Odonata is mostly subdivided into superfamilies (such as Aeshnoidea), families (Aeshnidae), subfamilies (Aeshninae), and tribes (Aeshnini). Each example is based on the genus *Aeshna*, defined by its type species, the Brown Hawker (*A. grandis*), characterized by its holotype (p. 25).

While such clear hierarchical language can elegantly express how species are related, it also conceals any uncertainty that may exist. A book about the diversity of insects that are still quite poorly known must therefore briefly address how the species are classified.

Ideally, a group name (as for a genus or family) covers all living descendants of a distinct lineage. Who belongs to which one is inferred from shared characters. As genus names ending in *neura* or *phlebia* (Greek for "nerve" and "vein," respectively) attest, wing venation details have been especially popular. These often strongly reflect species' shared ecology, however, being less informative of ancestry than genetics or features of the genitalia and nymphs (p. 41). Consequently, multiple groups that are not directly related were sometimes united under a single name (e.g. pp. 155 and 194). More frequently, one or more species from within a larger lineage were separated under a different name, because, for example, a minor venation character was given great weight.

As these taxonomic ranks are categories of convenience, practical considerations matter too. Changing as few genus–species combinations as possible, for example, disrupts the language that users are familiar with the least. Uniting hundreds of species in one genus, however, weakens that language's discerning power.

While odonate families are generally well-defined, fine-grained analysis of most

RIGHT | The wings have been much-used to group odonate species into genera and families, as the venation is complex and easily quantified. As the subtle asymmetries here show, however, the veins can also be too varied to be reliable!

relationships within them has yet to be done. There's enough evidence to know that many large and familiar genera do not reflect the shared and exclusive ancestry of the species implied, but too little evidence to rearrange them with confidence (see profiles on *Libellula*, *Pseudagrion*, *Rhinocypha*, and *Drepanosticta*). Meanwhile, new genera are proposed for species whose relationships have not been tested at all.

Relationships between many families are also poorly understood, but there the evidence may simply have eroded with time. Fossils are not just rare but consist largely of fragments of (thus potentially misleading) wings. Most of the genetic code, meanwhile, is either extremely conserved or has been endlessly jumbled and rewritten, making informative sections hard to find among millions of base-pairs of DNA.

IDENTIFYING ODONATES

Like birds, odonate adults are visual beings that recognize each other by differences we perceive instinctively too (p. 12), most species being distinct enough to identify from a photograph. Like their vertebrate counterparts, however, individual variation in the "birder's insects" can be confusing. Aside from marked sexual dimorphism (pp. 160–1, for example), colors change considerably with age. Many species are also flexible ecologically, giving the nymphs' habitat much influence over adult appearance, particularly their size and the extent of black markings. Wing venation is much used in identification but an abundance of veins also allows for considerable variation (see photo p. 26).

When diagnosing multiple species together, for instance as genera or families, this variability multiplies (see also p. 196). The easiest characters to see, such as male colors and claspers, moreover, typically evolve (and thus differ) at the species level. Traditional identification tools, such as dichotomous keys that work from families, through genera, and down to species, may thus be challenging. Fortunately, bottom-up approaches to separate the scores of species found together more visually, such as photo guides, are appearing for more and more regions.

Probably, adult identification will become increasingly organic: photos are classified using automated image recognition, validated within social networks of expert-enthusiasts, and, where necessary, verified by more technical means, such as the detailed examination of venation and genitalia with which odonate study began (see photo p. 22). This development rests heavily on observers amassing photographic material and field experience, capturing the species' beauty but their often substantial variability too.

ASSESSING ODONATES

Field recorders can also help monitor how odonates and their habitats are faring. In many parts of Europe, where the fairly poor fauna poses fewer identification challenges and their study also has the longest history, most data to assess species' occurrence and even population trends are assembled by citizen scientists on platforms such as Observation.org.

Such data may feed into the assessments for the International Union for the Conservation of Nature (IUCN) Red List of Threatened Species, the foremost tool for gauging the state of global biodiversity. Odonata are the first major insect group to be completely evaluated, with 16 percent of species estimated to be at risk of extinction. This seems low compared with the less mobile amphibians (41 percent) or often directly persecuted mammals (26 percent), but as data are insufficient to assess 29 percent of the world's species with confidence, 41 percent might actually be in danger. Habitat destruction is the foremost threat, forest stream species being most at risk (e.g. p. 234).

A more refined environmental assessment tool such as the Dragonfly Biotic Index gives a 0 to 9 score to each species, with up to 3 points awarded for the population trend (based on IUCN Red List classification) and ecological sensitivity (using a measure of habitat specificity, for example) of every one, as well as up to 3 for the unique biodiversity they represent, expressed by the relative extent of their world range.

LEFT | Both sexes of the Variable Sentinel (*Orchithemis pulcherrima*) from southeastern Asia are largely orange when fresh. While the male (middle photo; note secondary genitalia and claspers) can attain a bright red abdomen, it may also turn largely black with a white-pruinose saddle, quite like the mature female shown (bottom photo). Some females are wholly dark, though, while that shown at the top is still darkening.

RIGHT | Most threatened odonates are specialists with small ranges. A third of *Megalagrion* species (p. 140) risks extinction, for example, but this Hawaii Upland Pinapinao (*M. hawaiiense*) still seems safe.

ODONATA AS CONSERVATION ICONS

While "only" 16 percent of odonates may be threatened, 24 percent of freshwater species assessed are at risk overall, much more than in the sea and probably on land too. Obviously, protecting life in rivers that run through our cities, fill our dams, and irrigate our fields is even harder than setting aside some land or sea. This mostly small and submerged life is easier to neglect too.

Being so connected, freshwater catchments are the optimal units for managing resources on land for both humans and nature, however, and to protect the right to life in general, as embraced by the Rights of Rivers movement. Similarly, odonates are ideal proxies for overall non-marine biodiversity, because all species straddle water, land, and air.

When IUCN launched the Green Status of recovering species in 2021, meant to balance the fundamentally negative framing of its Red List—and, inevitably, most conservation messaging—it was therefore appropriate that the only species profiled to profit from general environmental legislation, rather than species-focused action, was a dragonfly (see p. 92).

Focusing on aquatic ecosystems and their flying inhabitants implies a wider view on conservation, integrated with environmentalism. While odonate diversity can be linked to many of today's foremost concerns (climate change, water scarcity, desertification, pollution, urban health), many means to do so are not widely available yet. The greatest contribution to their conservation, therefore, is to facilitate, through their popularization, a more biodiversity-inclusive approach to protecting the environment:

Help others identify the species. Platforms such as iNaturalist can only deliver their promise (informing conservation, automated photo identification) if records are vetted much more thoroughly. Luckily, anyone can gain that

FAR LEFT | Found only in vital freshwater and yet almost anywhere, odonates are symbols for a healthy planet, *Azuragrion granti* being confined to the few permanent pools on the arid island Socotra.

LEFT | As these fruity shandies from South Africa named after the Cherry-eye Sprite (*Pseudagrion sublacteum*) and Ruby Jewel (*Chlorocypha consueta*) demonstrate, coining appealing names can popularize odonates in unforeseen ways.

RIGHT | This unique *Rhyothemis* species remained undocumented due to decades of war in central Angola, an ecologically fragile and thinly populated region harboring the sources of the Congo, Zambezi, Okavango, and Cuanza. The Chokwe people believe those rivers will run dry if the water spirit Mukisi Nkisi leaves them: the new dragonfly will therefore be named in its honor.

experience. Most field guides are written simply because the expert-enthusiast author could not find what they needed. Every reader of this book can become an author.

Document ecology and behavior. Many questions posed in this book can be answered by observing nymphs and adults, or with simple experiments in the field or aquarium. Global collaboration on the cosmopolitan *Pantala flavescens*, for example, could clarify the weather's impact on them and their impact on mosquito numbers, giving a common goal to (citizen) scientists worldwide (p. 53).

Create habitat where there is none. Digging a pond might only benefit species that are already common, but there's no better way to introduce the dynamism, color, and behavior of dragons and damsels into your life and the lives of those around you. It is the ultimate educational tool.

COINING COMMON NAMES

The very first step in promoting odonates is to literally integrate them into our language further. Names in English, which tend to be popular with birders and naturalists around the world, are emerging rapidly. Today, we can encounter dragonlets and demoiselles, spreadwings and threadtails, emeralds and malachites, fineliners and parasols, pondhawks and billabongflies, treehuggers and riverkings, titans and pixies, and darters, darners, dancers, and dashers.

Some 1,000 species have even received multiple English names. A fitting name in Britain might not work in Europe, or one from Hong Kong may not be suitable in India or Australia. As long as language is understood locally, there's little harm in duplication. As the love for odonates spreads, consensus should increase, as it has in birds, while names in other languages will become more prevalent too. And, as everyone uses language, anyone can contribute.

DIVERSITY OF THE WORLD'S DRAGONFLIES AND DAMSELFLIES

The profiles contained in this directory characterize all major groups of Odonata. The sequence reflects our current understanding of relationships (but see below), with larger and more familiar groups treated first so more obscure ones follow later. Typically, taxonomists place the branches of the family tree with most living descendants last, so to some readers the family order will seem to be flipped! Such a sequence does not serve the book's purpose, however, as it automatically brings rather disparate and localized groups (such as kiwis and ostriches in birds) forward, while moving more widespread and better-known groups (parrots, songbirds) to the back.

Each profile is self-contained and yet serves to reveal how the overall diversity of odonates arose (or was interpreted by curious humans), so numerous references to other sections are provided throughout. While 46 families are recognized, these vary from six with just one species to three with over a thousand, so they have been broken up into more manageable and mostly natural chunks, based on the latest evolutionary studies, supplemented where necessary with the author's expert judgment.

Many of the 690 genera recognized, moreover, are still poorly defined (p. 27). As mentioning them without comment (or omitting them completely) would create a false impression of our understanding of overall diversity (in a book intended to introduce it!), at least some context is provided for every genus.

Fortunately, extensive genomic research is underway, so we can expect the species' allocation to families, subfamilies, tribes, and

genera to vastly improve in the coming years. The isolated genera that were removed from Corduliidae before they could be placed in other families (p. 84), for example, may well be reclassified as this book comes out. Updates can be found on the World Odonata List at www.odonatacentral.org.

Diagnosing groups of species morphologically can be technical and complex and, in such a compact book as this, thus quite confusing and potentially even misleading (p. 28). While a description of each family's adult habit is provided, therefore, the book should be seen as a guide to understanding odonate diversity, not to identifying it.

ABOVE | While almost any freshwater has odonates, virtually all of over 6,400 species breed only there. Morphing from aquatic nymphs to airborne adults, species like this emerging Common Goldenring (*Cordulegaster boltonii*) in Europe (left), displaying Green Metalwing (*Neurobasis chinensis*) in Asia (middle), and resting Strange Helicopter (*Anomisma abnorme*) in South America (right) thus bring a vital resource and environment, as well as freshwater's perpetual cycle of renewal, to life in symphonies of color.

Indeed, the profiles tend to focus on features that help understand the groups' ecology and evolution, such as the adult females' egg-laying scoops or nymphs' digging legs. Similarly, size is variable and uninformative in most contexts, so rough measurements are mostly just provided where they serve the general understanding of odonates—for example, for exceptionally large or small species.

Information on behavior and ecology, finally, is also presented mostly within the context of the groups' distribution, diversity, and history.

Key to the maps: Each group's distribution is shown in pink on a world map, with darker and blue shading (and purple where there is overlap) sometimes used to indicate where specific genera occur. This use of shading is explained where appropriate in the text or panel. Where no legend is provided, the pink area thus shows the entire group's range with the exception of those genera indicated by other shades.

ANISOPTERA

FAMILY: LIBELLULIDAE

SKIMMERS

The first dragonfly most people are likely to see is one of over 1,030 species placed in Libellulidae. The family is so common and diverse because its species deal best with change, potentially breeding in any freshwater, even when it is short-lived or polluted. Adults are often strong fliers, moreover, finding homes over great distances, even deep in deserts and across oceans.

ABOVE | While not closely related, the male of Asia's Red Sultan (*Camacinia gigantea*) appears like a gigantic, hyperactive parasol (*Neurothemis*).

RIGHT | Madagascar Riverking (*Zygonoides lachesis*) female pressing out eggs. Like most libellulids, she will lay these by rapidly tapping the water surface.

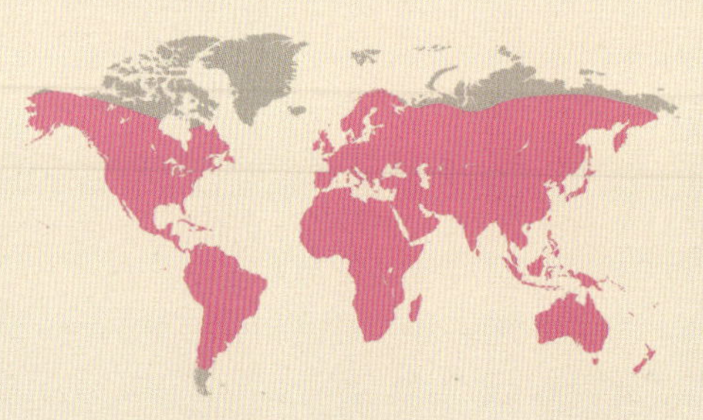

DIVERSITY
1,035 species that may appear at almost any water where odonates occur

ADULT HABIT
Extremely varied, from the tiniest to the bulkiest odonates; most perching dragonflies with bright colors (such as red and blue-pruinose), especially at standing waters, belong to this family

TAXONOMY
At least 11 fairly well-defined subfamilies are recognized, but their further subdivision is unsettled and the position of some genera uncertain (p. 43), so the larger subfamilies are spread across several profiles based on their ecology

Once suitable habitat has been found adults tend to perch; yet they remain among the most active odonates, constantly flying off to engage each other or prey and then returning to their post. This busy and often exposed lifestyle means they must manage their temperature carefully, so they often posture their body and wings to maximize or minimize the sun's impact (p. 54).

The perching habit coincides with the greatest diversity in appearance of any odonate family, ranging from the portliest species to the most petite, smaller even than most damselflies. Libellulids also cover the full spectrum of odonate colors, although shades suited to sunny places that also allow males to be seen by their mates and rivals (and fortunate onlookers such as ourselves) predominate.

Green hues, so common as camouflage in insects, are scarce. Blue pigments, dominant in other families (pp. 118 and 148), are even exceedingly rare (p. 49). Red pigments and pruinosity (reflective wax appearing white to gray or blue), however, are applied to exceptionally great effect, combining as dazzling violet on the body, for instance, or as bright wing flashes. Indeed, no other dragonfly family so often has colored and extravagantly patterned wings.

Combining great adaptability with strong dispersal and diverse reproductive behaviors, libellulids thus became the ultimate dragonflies. Frequently, they will have replaced other groups, perhaps in direct competition (p. 75) but also by responding faster whenever Earth's endless cycles of change impacted existing habitats and created new ones (see p. 20).

LIBELLULIDAE—LIBELLULINAE—*ORTHETRUM* AND *ORTHEMIS*

TYPICAL SKIMMERS

Found at virtually any exposed stagnant water in the hot parts of Asia, from fishponds and rice paddies to filthy city drains, the Slender Skimmer (*Orthetrum sabina*) is an exemplar of the "superdragons" described in the family introduction, extending far across the islands of the Pacific and the drylands of Africa as well.

This and the over 270 other species of Libellulinae, the largest subfamily of Libellulidae, may owe their particularly substantial contribution to the family's overall success to relying less on structure provided by plants (aquatic and riparian vegetation, roots, leaf-litter, wood), which allows them to take up opportunities also where plants

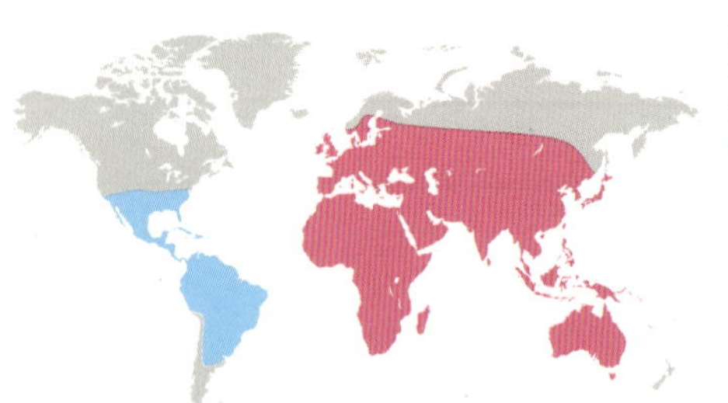

DIVERSITY

94 species found at almost any freshwater, especially in more open landscapes and warmer regions

TAXONOMY

Genera *Orthetrum* (66 species; **pink on map**) and *Orthemis* (28; **blue**) have similar ecologies and names (p. 39) but are probably not each other's closest relatives

are struggling or just beginning to grow. Most nymphs are suited to life in soft water-bottoms, being sturdy with strong legs to bury themselves in mud and silt. Their square heads bear small eyes that can stick up to peek above the gunk. The body is generally covered densely with hairlike setae, concealing the nymph by holding sand or debris, but probably also helping it to feel its way around in the soupy substrate.

With few exceptions, females have distinctive side flaps where the eggs exit the abdomen tip. Many scoop water from the surface and flick egg-laden drops onto the bare bank, some bulky species of rainforest puddles (p. 41) propelling them over a meter away. Presumably, the eggs are safer there from predators, but close enough for the hatched nymphs to quickly reach water.

While used generally for members of the Libellulidae (p. 34), the name "skimmer" refers more strictly to the common Libellulinae of open water. Indeed, flying low over ponds and lakes all over Europe, the Black-tailed Skimmer (*Orthetrum cancellatum*) first carried that name. Among these typical skimmers, the genus *Orthetrum* is most diverse, occurring throughout the Old World from forest streams and desert pools to bogs and saltmarsh. The tropical king skimmers (*Orthemis*) are their ecological counterparts in the warmer parts of the New World, the king skimmers (*Libellula*) in the temperate north (p. 38).

LEFT | The aggressive Slender Skimmer (*Orthetrum sabina*) habitually eats other dragonflies, including its own species.

TOP RIGHT | Although some Asian species are bright red, most mature males of *Orthetrum*, such as this Bold Skimmer (*O. stemmale*) hovering over a pool in Ghana, are extensively pruinose.

RIGHT | Yellow-lined Skimmer (*Orthemis biolleyi*) from Central America showing the side flaps near the abdomen tip that are characteristic of females in the subfamily Libellulinae.

TEMPERATE SKIMMERS

Like most odonate groups, this subfamily diversified largely in the tropics (p. 40), only *Libellula* doing so in the temperate north instead, although the exceptionally bulky Hercules Skimmer (*L. herculea*) has a wide range in tropical lowlands, inhabiting forest pools in Central and South America. Unsurprisingly therefore, this dragonfly genus was the first (in 1758) to be named by Carolus Linnaeus, Sweden's founder of modern taxonomy. *Libellula* is a diminutive of *libella*, the Latin word for "mason's level," which is T-shaped like a dragonfly (*libella* is itself diminutive of *libra*, a balancing scale). *Libella* remains the origin of the word for odonate in many European languages.

Most genus names in Libellulidae end in *themis*, Greek for "custom" or "law," Themis being the goddess of justice and divine order. That sounds poetic, but Hermann Hagen never explained his intentions. Working on the first treatise of North American Odonata, the German was probably inspired by some rather prosaic names coined by Edward Newman in 1833, three decades earlier. The skimmers there (p. 37) reminded Hagen of Newman's *Orthetrum* (Greek for "straight body")

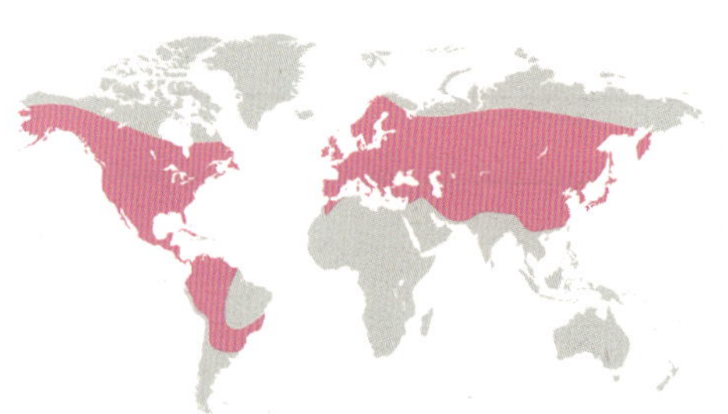

DIVERSITY
33 species at most standing and slow-moving waters in Northern Hemisphere's temperate zone, with one widespread on forest pools in American tropics

TAXONOMY
Genera *Ladona* (3 species), *Libellula* (28), and *Plathemis* (2)

TOP LEFT | Broad-bodied Chaser (*Libellula depressa*) male from Germany. Together with the Four-spotted Chaser (*Libellula quadrimaculata*) to the right, this species and genus were the first dragonflies to be named scientifically.

LOWER LEFT | A male of the Neon Skimmer (*Libellula croceipennis*), which is widespread in both North and Middle America.

RIGHT | Four-spotted Chasers (*Libellula quadrimaculata*) roosting in the United Kingdom.

and two whitetail species were reminiscent of Europe's Broad-bodied Chaser (*Libellula depressa*), called *Platetrum* ("broad body") by Newman, hence *Orthemis* and *Plathemis*. *Themis* thus came to mean "libellulid-like dragonfly," so two-fifths of the family's genera have it in their name today. With pronounced genitalia, *Orchithemis* and *Pornothemis* are "testicle-libellulid" and "prostitute-libellulid," for example (p. 42).

Ironically, a "divine order" in naming Odonata is still elusive (p. 26). Cynthia Longfield was so unsure about naming yet another similar group in 1955 that she called her African genus *Nesciothemis*, "I do not know-libellulid" (p. 42)!

The disorder, moreover, goes back right to the start. Besides *L. depressa*, Linnaeus named *L. quadrimaculata*, called Four-spotted Chaser in Europe and Four-spotted Skimmer in North America. Four other chasers from Eurasia and 27 king skimmers from the Americas were subsequently placed in *Libellula*, although these are so varied that a slew of additional genera has been proposed, such as *Belonia* and *Holotania*.

Only *Plathemis* and *Ladona* for North America's whitetails and corporals are often used, although none of these species' affinities have been fully resolved, particularly relative to the many tropical genera (p. 40). The ultimate stumbling block that remains unresolved seems particularly inane: which of Linnaeus's two species actually gets to keep the genus name?

JUNGLESKIMMERS, GRENADIERS, AND ASSOCIATES

While the subfamily Libellulinae's dominance at open and disturbed habitats may have been facilitated by a reduced dependence on plant matter (p. 36), its diversity appears to have originated in lush tropical climes. Between these extremes, a staggering (but confusing!) variety of sizes, shapes, and colors evolved.

Retaining water in the smallest spaces, rainforests abound in potential habitat: even the squelching forest floor may provide

ABOVE | A male of the Common Grenadier (*Agrionoptera insignis*) on Sulawesi in Indonesia.

RIGHT | Madagascar Jungleskimmer (*Thermorthemis madagascariensis*) female laying eggs by propelling them in drops onto the bank.

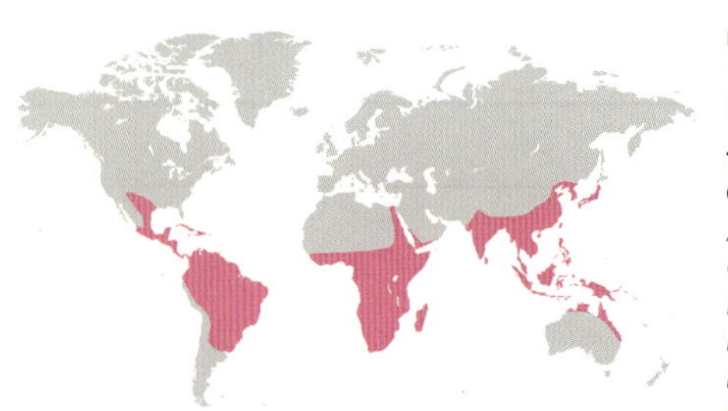

DIVERSITY
147 species at almost any water in the tropics where odonates occur

TAXONOMY
Genera *Aethiothemis* (13 species), *Agrionoptera* (8), *Amphithemis* (3), *Cannaphila* (3), *Cratilla* (2), *Dasythemis* (4), *Diplacina* (28), *Epithemis* (2), *Hadrothemis* (7), *Hylaeothemis* (4), *Hypothemis* (1), *Lathrecista* (1), *Lyriothemis* (19, including *Boninthemis*), *Micromacromia* (4), *Misagria* (4), *Neodythemis* (14), *Nesciothemis* (5), *Nesoxenia* (2), *Notolibellula* (1), *Orchithemis* (3), *Oxythemis* (1), *Pacificothemis* (1), *Palaeothemis* (1), *Phyllothemis* (2), *Pornothemis* (2), *Potamarcha* (2), *Protorthemis* (4), *Pseudagrionoptera* (1), *Tapeinothemis* (1), and *Thermorthemis* (2). All (probably) belong to Libellulinae, but position of *Akrothemis* (2) unconfirmed

a place to breed. To profit from the limited resources in such constricted habitats, some species are tiny, while others may be very large in order to compete (compare p. 156).

Throughout the tropics, for example, the puddles mushing up the forest roads are guarded by big fat libellulines, such as the Malagasy jungleskimmers (*Thermorthemis*), *Protorthemis* in New Guinea and Sulawesi, and the Americas' burliest *Libellula* and *Orthemis* (pp. 37–8). Most bombardiers (*Lyriothemis*) in Asia and the African jungleskimmers (*Hadrothemis*) also breed in muddy pools, but some favor water-filled cavities in trees.

In such wet environments, suitable habitat is typically closeby, so many species adapted with narrow wings, suitable for taking off quickly into the canopy but not for going far. Consequently, their veins were often rearranged and reduced, inspiring the description of numerous genera (p. 26). The name *Agrionoptera*, for example, means damselfly-wing. Genetics show this genus should not only include the grenadiers found at leafy pools and forest swamps from tropical Asia to the Pacific, however, but also the bulky *Protorthemis* and the smaller Bicolored Skimmer (*Notolibellula bicolor*) from north Australia's rock pools. In Asia, the *Nesoxenia* grenadiers and Bloodtail (*Lathrecista asiatica*) are also related, as may be the pursuers (*Potamarcha*) and forestskimmers (*Cratilla*). Even the structurally and ecologically similar *Cannaphila*, *Misagria*, and especially *Dasythemis* might well have been placed in *Agrionoptera* if they were not from Central and South America! Forest libellulines everywhere have the piebald thorax stripes for which those genera have been named "convict skimmers."

To absorb the reduced sunlight in their shady environs, or to blend in, adults can also be largely black, marked only with dabs of yellow or green. Larger spots frequently lie near the abdomen tip, perhaps serving as signals. Such unshowy males may need other ways to woo females, perhaps with elaborate claspers or genitalia (compare p. 99).

From the study of genitalia, nymphs, and genetics, it has become clear that some libellulids have changed so much that they are barely still recognizable as such. Earlier odonatists considered the smallest and darkest species with the narrowest wings to be more primitive, like an ancestral group from which larger and more colorful ones evolved (p. 72). The opposite, however, appears to be true. The claspers of some junglewatchers (*Hylaeothemis*, *Neodythemis*) and micmacs (*Micromacromia*), small dragonflies inhabiting forest streams and seeps in tropical Asia, Africa, and Madagascar, for example, put the similarly black-and-yellow Gomphidae to shame (p. 92). The stream-living *Diplacina* from the Philippines, Sulawesi, the Moluccas, and New Guinea, as well as the streamwatchers (*Phyllothemis*) in Southeast Asia, are similarly disparate.

Sometimes only the female's sideflaps (p. 37), or details of nymphs or genitalia, reveal that species are part of Libellulinae. Uncovering their exact affinities to infer their history and ecology, and classify them correctly, is even harder. In tropical Africa, the flashers (*Aethiothemis*) have a distinctively exposed penis in common, but were initially described in five different genera. Longfield's confusing blacktails and peppertails (*Nesciothemis*) proved close to the Pepperpants (*Oxythemis phoenicosceles*) (p. 39).

Much work remains to be done. Restricted to Sundaland's swamp forests and mangroves, the sentinels (*Orchithemis*; photos p. 28) and marshals (*Pornothemis*) recall miniature and thin-bodied *Lyriothemis*. The similarly small hawklets from India's Western Ghats (*Epithemis*) and Indochina (*Amphithemis*) appear close too.

Some of the most modified genera are so poorly known that identifying their true affinities is still quite impossible. Fiji's *Hypothemis*, Pohnpei's *Pacificothemis*, Myanmar's *Palaeothemis*, Borneo's *Pseudagrionoptera*, and the Solomon Islands' *Tapeinothemis*, all with a single species, do appear to be near Libellulinae. Two species of *Akrothemis* from New Guinea, however, might in fact belong to one of the groups treated next.

ABOVE LEFT | This male Spring Micmac (*Micromacromia zygoptera*) from Ghana is one of many inconspicuous libellulids confined to rainforest streams.

ABOVE MIDDLE | The male Pepperpants (*Oxythemis phoenicosceles*) from tropical Africa shows off his brightly colored legs.

ABOVE RIGHT | This male *Diplacina phoebe* from Halmahera, Indonesia, and its relatives from New Guinea are likely to form a separate genus from the true *Diplacina* species found on Sulawesi and the Philippines.

RIGHT | Red-rumped Hawklet (*Epithemis wayanadensis*) male. Only the second of its genus, this southern Indian species was first described in 2023.

LIBELLULIDAE—PALPOPLEURINAE—*CROCOTHEMIS* AND RELATED GENERA

SCARLETS, DRAGONLETS, AND KIN

Humans have moved many insect species around the globe, often with dire consequences. The only truly invasive odonate, however, is the Oriental Scarlet (*Crocothemis servilia*); besides Hawaii (where five species introduced from North America have become established), this Asian species gained a foothold in Florida in the 1970s and on Cuba in the 1990s, and now occurs on all islands of the Greater Antilles. There is no trade and thus no deliberate transport of odonates. As odonates often disperse well, there may be fewer unoccupied niches available for invasion than is the case for other insects. So how did this species succeed?

Crocothemis belongs to Palpopleurinae, the second-largest subfamily in Libellulidae, with almost 170 species currently classified. Like Libellulinae (p. 36), they are abundant and conspicuous at most standing, open, warm waters, which are often recent or seasonal, forming after the ground has been disturbed or rain has fallen.

LEFT | Africa's most widespread dragonfly, the Broad Scarlet (*Crocothemis erythraea*), is superficially identical to its Asian counterpart, the Oriental Scarlet (*C. servilia*).

RIGHT | This female Indian Rockdweller (*Bradinopyga geminata*) has excellent camouflage.

DIVERSITY
118 species in a wide range of warmer waters (mostly open and standing, often temporary) around the world

TAXONOMY
Genera *Bradinopyga* (4 species), *Crocothemis* (9), *Diplacodes* (10), *Erythrodiplax* (61), *Hemistigma* (2), *Indothemis* (2), *Nannodiplax* (1), *Neurothemis* (17), *Palpopleura* (7), and *Thermochoria* (2) appear close, but *Anatya* (2) and *Pseudoleon* (1) treated here somewhat tentatively

Unlike libellulines, most palpopleurine nymphs are smooth with large eyes and slender legs, indicating they roam freely rather than burrow, and appear to lack dorsal abdominal spines. While these can be absent in other subfamilies too, most libellulid genera and their presumed ancestors have them. As spines protect against fish, their consistent absence in such a dominant group suggests that their nymphs manage best in small, isolated, ephemeral waters without fish, or in microhabitats fish cannot penetrate.

Rockdweller (*Bradinopyga*) nymphs, for example, amble about fearlessly in rock pools (and cement basins and water tanks!) in Africa and Asia. The speckled adults sit flat on rockfaces and walls nearby, blending in completely. Nymphs of most palpopleurines clamber among submerged plants, however. The females place their eggs onto these plants, organic matter or mud, using the genital opening's extended (and often downcurved) lip (compare Libellulinae; p. 37).

Scarlets and rockdwellers are among almost 60 such species from the Old World's warmer reaches. Found at any rice paddy or roadside pool, their often colorful wings mean they may be the most photographed odonates: in Africa, perhaps the black-winged Lucia Widow (*Palpopleura lucia*) wins; from Asia across to Australia, one of the parasols (*Neurothemis*) with their black-and-white or largely

RIGHT | This male Lesser Red Parasol (*Neurothemis fluctuans*) from Thailand must be agitated, as most parasols hold their colorful wings down, shading the body like an umbrella.

red wings. The perchers (*Diplacodes*) too are very numerous on these continents.

Also in the warmer parts of the New World, about 60 species are among the most-seen dragonflies. The majority are classified as dragonlets (*Erythrodiplax*), which have diversified into an even wider array of habitats: Costa Rica's Canopy Dragonlet (*E. laselva*) breeds in bromeliads holding water high in rainforest trees; the Seaside Dragonlet (*E. berenice*) favors brackish water in mangroves, saltmarsh, and deserts.

The tropical distribution and success of the subfamily Palpopleurinae suggests that only cold winters limit its conquests. The shallow and well-vegetated waters they favor might warm up in summer, but leave the nymphs exposed when temperatures drop. Nonetheless, all Africa's most widespread odonate, the Broad Scarlet (*C. erythraea*), needed to expand across Europe was some eutrophic ponds and a bit of global heating.

Perhaps all its near-identical Asian counterpart needed in America, therefore, was a head start. The Oriental Scarlet is one of the commonest species in tropical Asia, where many aquarium and pond plants are cultivated for export. Unsurprisingly, therefore, it is the dragonfly imported accidentally most often, of 40 odonate species reported so far (see also p. 136).

The ecologically varied but closely related species of this subfamily have often converged

in appearance: separating some *Diplacodes* and *Erythrodiplax* is hard without knowing the continent they are from. Others look more distinct than they are. Despite its grizzled body, tiger-striped eyes, and blotched wings, the Filigree Skimmer (*Pseudoleon superbus*), adapted to rocky streams in Middle America, is near *Erythrodiplax*, for example. Two blue-eyes (*Anatya*) from rainforest pools seem near that genus too.

Many genera are ill-defined, therefore. The piedspots (*Hemistigma*) and piedfaces (*Thermochoria*) from Africa and Madagascar differ only in their veins' density. One of six widows (*Palpopleura*) there (the seventh is Asian) looks so deceptively like the former that it was once named *Hemistigmoides deceptor*! The Pygmy Percher (*Nannodiplax rubra*) from Australasia is just a tiny *Diplacodes*, while tropical Asia's two demons (*Indothemis*) look like rather robust, black examples of that genus.

LEFT | This female Asian Widow (*Palpopleura sexmaculata*) strongly recalls the Americas' distantly related amberwings (p. 66).

TOP RIGHT | The male of Middle America's unique Filigree Skimmer (*Pseudoleon superbus*).

LOWER RIGHT | Band-winged Dragonlet (*Erythrodiplax umbrata*) male from the Cayman Islands.

PONDHAWKS, PINTAILS, PYGMYFLIES, AND KIN

The bright colors of the male Bluebolt (*Cyanothemis simpsoni*) in Africa, Greenbolt (*Viridithemis viridula*) on Madagascar, and redbolts (*Rhodothemis*) from south Asia to Australia inspired both their common and scientific names (p. 38). Their American counterparts are just as colorful: the redskimmers (*Rhodopygia*) are a particularly vivid red, while some pondhawks (*Erythemis*) are red or black and others largely green. Like many libellulids, pruinosity can affect their look too, with age turning two green *Erythemis* species wholly blue and a red *Rhodopygia* purple; Brazil's *Carajathemis simone* is pruinose with a red tip.

RIGHT | A female Great Pondhawk (*Erythemis vesiculosa*) in Suriname feeding on a close relative, a male Black Pondhawk (*E. attala*).

LEFT | The weirdly shaped pintails (*Acisoma*), such as this male Stout Pintail (*A. inflatum*) from Ghana, are found within grassy verges from tropical Africa to Madagascar and Asia.

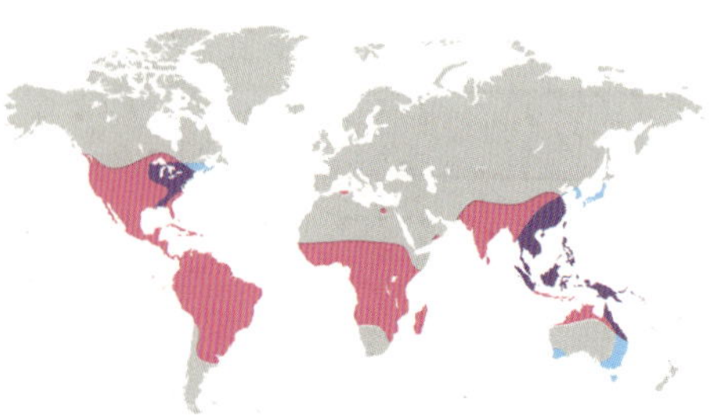

DIVERSITY
43 species of richly vegetated (standing and sometimes running) waters; almost global, but absent from much of Eurasia

TAXONOMY
Genera *Nannophya* (8 species) and *Nannothemis* (1), **both blue on map**, appear closer to each other than to *Acisoma* (6), *Carajathemis* (1), *Cyanothemis* (1), *Erythemis* (10), *Porpax* (6), *Rhodopygia* (5), *Rhodothemis* (4), and *Viridithemis* (1) **in pink**

Representing only 43 species of the dragonflies' largest and gaudiest family (p. 35), this group is indeed unrivaled in color range. The Bluebolt and some pintails (*Acisoma*) and pricklylegs (*Porpax*) in Africa are the only libellulids whose deep blue markings are not caused by pruinosity but, as damselflies (p. 148) and aeshnid dragonflies (p. 112), by pigments. And while many dragons have touches of green, notably those in dark habitats, that color rarely covers the entire body of those sitting out in the sun; its red tail-end makes the Greenbolt male look like a popsicle!

Remarkably, the group's size range is unequaled too. The 2⅓ in (6 cm) Great Pondhawk (*E. vesiculosa*) habitually takes butterflies and other dragonflies as prey. While the Bluebolt is almost as big, its cousins in the genus *Porpax* include Africa's smallest dragonfly, less than ¾ in (2 cm) long. Down to ⅗ in (15 mm), the pygmyflies (*Nannophya*) from Japan to Tasmania are even the shortest odonates. Their North American sister, the Elfin Skimmer (*Nannothemis bella*), is almost as tiny.

All species favor waters rich in vegetable matter, such as swamps and pools with lush vegetation or

thick deposits of leaf-litter, rivers with floating organic debris, or seeps with wet cushions of peatmoss. Males perch persistently near preferred microhabitats, often defending them vigorously. Deep within grassy borders, pygmyfly territories are under a square meter in size. Strong evolutionary selection on size and color either led them to fit into such specific niches, therefore, or rather stand out from them. Catching big prey might even have evolved from aggressive territoriality, literally eating the competition!

Many species press their body, wings, and legs against leaves or the ground when perched. Perhaps that is why their bright colors are often concentrated dorsally and the occiput (the triangle separating the top of the eyes) is diagnostically large: to make them more visible. Large spines and dense setae on the hind femora, especially in some males, may be linked to this posture too. These can aid the capture of large prey but, judging from their diversity in pricklylegs, also have a sexual function.

ABOVE | This male Scarlet Pygmyfly (*Nannophya pygmaea*) from east Asia may be the world's tiniest dragonfly.

BELOW | While many libellulids appear blue, the male of Africa's Bluebolt (*Cyanothemis simpsoni*) is among very few species where this color is not produced by pruinosity.

WOODSKIMMERS

Many palpopleurines breed in seasonal waters, the adults sheltering in woodland for months and only attaining their mature colors as the first rains flood the landscape. In lowland forest in Central or South America, great numbers of dull dragonflies with dark-marked wings perch among the undergrowth just above the ground. Brown with a unique wood-grain pattern of fine transverse lines on the thorax, they eventually become dark and dusted with gray pruinosity.

Males of *Uracis* species, and the much smaller Little Woodskimmer (*Ypirangathemis calverti*), differ mostly in their wings, which are black-tipped in some, broadly banded in others. Alongside the black bands, the White-banded Woodskimmer (*U. siemensi*) develops pruinose flashes in the wing bases, a piebald pattern that stands out as it hovers over small shaded waterholes.

Female woodskimmers jab their eggs in flight into muddy pools, damp earth, and even dry soil in shallow depressions before those inundate. While drawn out in most Palpopleurinae (p. 45), the plate below the female's genital opening is developed most dramatically here, even projecting far beyond the abdomen like a stinger (thus recalling Cordulegastridae; p. 86) in the Spike-tailed Woodskimmer (*U. ovipositrix*).

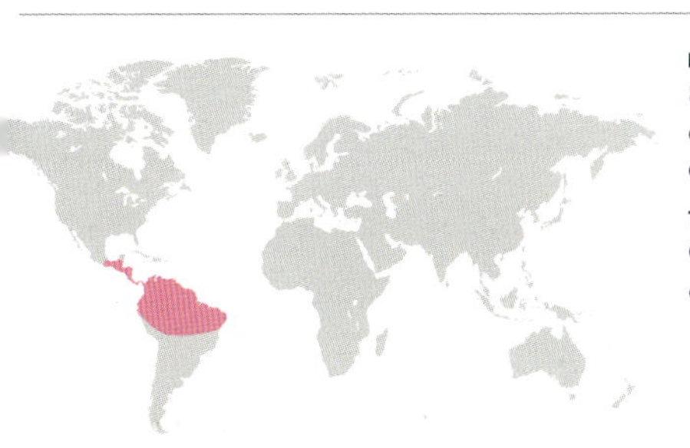

DIVERSITY
8 species of seasonal puddles on the forest floor in Central and South America

TAXONOMY
Genera *Uracis* (7 species) and *Ypirangathemis* (1)

TOP | A male of the Large Woodskimmer (*Uracis fastigiata*) in Panama.

ABOVE | Female of the Common Woodskimmer (*Uracis imbuta*) demonstrating the group's characteristic wood-grain pattern.

LIBELLULIDAE—PANTALINAE—*PANTALA*

RAINPOOL GLIDERS

BELOW | Wandering Gliders (*Pantala flavescens*) mating in flight in Nepal.

With *Pantala* roughly meaning "throughout-roaming" in Greek, the Wandering Glider (*P. flavescens*) is indeed one of the most widespread insect species and most remarkable odonates. Also called the Globe Skimmer, it is the only one to occur on every continent but Antarctica (rare in Europe), reach the sub-Antarctic island of Amsterdam, and breed on Rapa Nui (Easter Island), over 1,200 miles (2,000 km) from the nearest populations.

Turning orange as they mature, adults are built completely for flight: robust (1 ¾–2 in/4.5–5.0 cm long) with long pointed wings, the very broad-based hind pair being distinctly triangular; the abdomen is cylindrical and tapered, almost conical. Built for growth, the big nymphs are just as unmistakable. They are active hunters with prominent eyes and teeth, and long abdominal spines and legs, the latter peculiarly black-tipped, contrasting with the almost translucent body. Melanin provides sturdiness, perhaps another adaptation to their rapid lifestyle.

DIVERSITY
Two species of open temporary ponds almost everywhere except Antarctica and much of Europe

TAXONOMY
Genus *Pantala* (2 species)

New generations emerge from rain puddles (and other warm and unvegetated waters) within about five weeks, which are among the fastest developments in a relatively slow-developing insect order.

The species is thus optimized entirely for living in the wake of summer rains: as the seasons shift, each generation must move across the equator to find a new bounty of balmy food-filled pools. Wherever a monsoon climate prevails, numbers build up toward the rains, swarming around treetops and patrolling over swimming pools and shiny car roofs. Tending to glide and hover, during rare resting periods they typically hang inside vegetation.

Only in 2009 it was inferred that their annual arrival on the bone-dry Maldives is part of what may be the greatest event of insect dispersal. Masses appear to fly directly from India to Africa each year, benefiting from favorable winds to cross the ocean in as little as four days. To get back, however, their offspring must take the long way around, following the coast.

Although some Trameinae look similar (p. 68), *Pantala* appears to have no direct relatives (but see next pages), the darker and thicker-bodied Spot-winged Glider (*P. hymenaea*) being the only other species. Recognized by a dark blotch in its hind wing expansion, this species has a similar ecology as its global cousin but only extends from Canada to Argentina, and from the Caribbean to the Galapagos.

ABOVE | Distributed globally, the Wandering Glider (*Pantala flavescens*) is the Worldwide Dragonfly Association's mascot. As harbinger of the rains and monsoon mosquito oppressor, it is the perfect focus for global collaboration among odonate enthusiasts (p. 31).

OPPOSITE | Wandering Glider (*Pantala flavescens*) nymph.

LIBELLULIDAE—PANTALINAE—*ZYGONYX* AND *TRITHEMIS*

CASCADERS AND DROPWINGS

As that family dominates ephemeral ponds (p. 34), it makes sense that the Wandering Glider (*Pantala flavescens*)—the farthest-flying and perhaps fastest-breeding odonate—is a libellulid (p. 52). Most pantalines favor permanent streams and rivers, however, forming the family's largest running-water radiation. Cascaders (*Zygonyx*) guard their domain in sustained flight and rest away from water hanging in vegetation, just like their wandering relative, but their spiny nymphs cling to rocks like limpets in rapid currents, even of waterfalls (see photo p. 9) .

Most genera perch frequently, though, such as the dropwings (*Trithemis*) that occupied almost the full complement of freshwater habitats available to them. Genetics suggest that the oldest extant species arose in Africa when forests gave way to savanna 10 mya, but most appeared in the last

TOP LEFT | Many libellulids, such as this male of Africa's Red-veined Dropwing (*Trithemis arteriosa*), will point their abdomen at the sun and also often lower their wings to minimize exposure, behavior known as obelisking.

LEFT | Africa's Ringed Cascader (*Zygonyx torridus*) extends to India and southern Europe, patrolling over open streams.

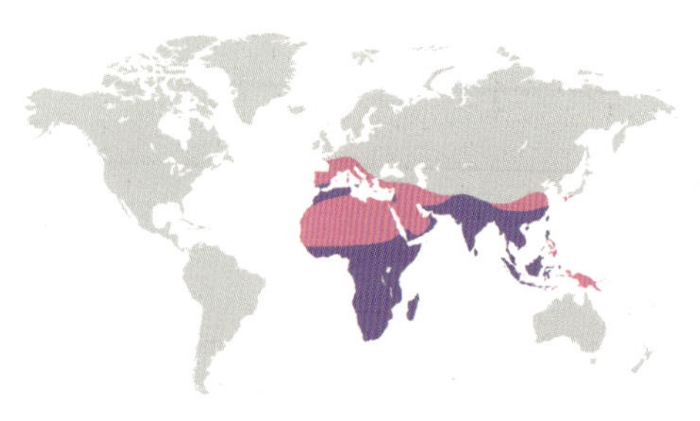

DIVERSITY

75 species mostly in wide range of flowing waters in tropics of Africa and Asia

TAXONOMY

Genera *Trithemis* (43 species including *Thalassothemis*; **both colors on map**) and *Zygonyx* (24; **purple**)

RIGHT | Appearing like a typical dropwing, *T. marchali* from the oceanic island of Mauritius was only separated from *Trithemis* in the genus *Thalassothemis* ("sea-libellulid") on account of its venation.

BELOW | This photo taken in Angola's highlands in 2018 was the first evidence of this dropwing's existence. Still unnamed, this new *Trithemis* species appears to mimic large *Acraea* butterflies.

5 mya. As the climate continued to fluctuate, they became isolated as habitats fragmented, or adapted as new ones appeared. Ultimately over 40 species thus evolved there, dominating dragonfly communities from cool streams to warm temporary pools, rainforests to deserts, and lowlands to highlands today.

Some species left Africa, resulting in another two on Madagascar and five in Asia, of which one got to New Guinea. And the exodus continues. The Violet Dropwing (*T. annulata*) arrived in Spain in the 1970s and is now halfway through France. Even inhabiting swimming pools and fountain basins, the Orange-winged Dropwing (*T. kirbyi*) crossed over only in 2008 but reached Belgium in 2022!

These species and most others of sunny habitats have red males (see photo p. 13). That color stands out against green plants but also against the open sky. To aid warming, most species in shadier places are largely black, often combined with blue pruinosity to still stand out. The species' wings, meanwhile, vary from clear and narrow with reduced venation at forest streams (compare p. 41) to broad with dense cells and butterfly-like blotches of color in open swamps (p. 64).

Many species were placed in separate genera to capture this diversity, but most have been sunk back into *Trithemis*. The next may be *Thalassothemis marchali*, restricted to rocky streams on Mauritius. Related to the Dancing Dropwing (*T. pallidinervis*), a south Asian colonist of new ponds, its ancestor must have crossed the Indian Ocean just as the Wandering Glider does annually today.

Indeed, the cascaders' restless habit and the dropwings' dramatic diversification show how such a species can evolve and colonize the world. On Rapa Nui, meanwhile, the glider has shorter wings and is said to fly lower, settling as soon as bad weather threatens to carry it offshore. Are these starved castaways or is the evolution of another island specialist like Mauritius's dropwing already underway?

RIVERDARTERS, STREAMSKIMMERS, AND KIN

Despite the extremes discussed in the previous profiles, most pantalines inhabit streams and rivers, often in forest. Each tropical region, for example, has robust species that fall ecologically between *Zygonyx* and *Trithemis*, with males aggressively defending stretches of rapid water from a conspicuous perch.

In Asia, these are the riverdarters (*Onychothemis*), in America, the streamskimmers (*Elasmothemis*) (**both marked pink on the map**). The related *Orionothemis* is known only from its extremely spiny nymphs found on Brazil's high plateau, reared to adult. Aside from *Pantala* (p. 52) these are the subfamily's only members to reach the New World.

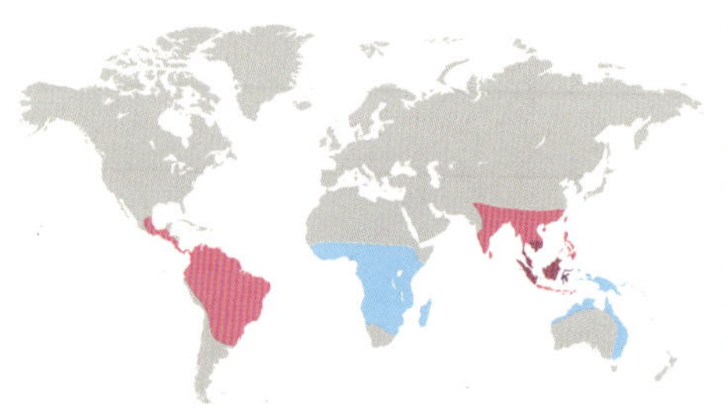

DIVERSITY

101 species of (often forested) running waters in the global tropics

TAXONOMY

Genera *Archaeophlebia* (1 species), *Atoconeura* (6), *Bironides* (5), *Celebophlebia* (2), *Celebothemis* (2), *Elasmothemis* (8), *Eleuthemis* (5), *Huonia* (15), *Lanthanusa* (7), *Malgassophlebia* (5), *Microtrigonia* (5), *Nannophlebia* (26), *Olpogastra* (1), *Onychothemis* (7), *Orionothemis* (1), and *Zygonoides* (4) included here, but position of *Risiophlebia* (2) is uncertain

In Africa and Madagascar, the riverkings (*Zygonoides*; photo p. 35) and Bottletail (*Olpogastra lugubris*) fill this niche (**blue on map**). With its yellow-studded black body and narrow but swollen-based abdomen, the latter is almost identical to *Celebothemis delecollei* from Sulawesi. The gap between them may seem surprising but fits the distribution of the many smaller stream-dwellers.

While *Celebophlebia* is also confined to Sulawesi, that genus appears related to the archtails (*Nannophlebia*) that occur from there to Australia, mostly on New Guinea (**blue and purple on map**). These tiny black-and-yellow species, moreover, seem close to another 32 limited to New Guinea, the adjacent Moluccas, and the extreme north of Australia, placed in *Bironides*, *Huonia*, *Lanthanusa*, and *Microtrigonia*. The highlanders (*Atoconeura*) are limited largely to Africa's tropical uplands and three leaftippers (*Malgassophlebia*; photo p. 12) and firebellies (*Eleuthemis*) to the lowlands, while the Furbelly (*Archaeophlebia martini*) and another two leaftippers inhabit Madagascar.

ABOVE | The archtails (*Nannophlebia*), such as this male Common Archtail (*N. risi*) from Australia, perch with the slender abdomen held curved.

LEFT | The aggressive male of the Bottletail (*Olpogastra lugubris*) in Ghana.

RIGHT | A male of the Blushing Streamskimmer (*Elasmothemis rufa*) from Suriname.

In Asia (**dark pink on map**) only the potbellies (*Risiophlebia*) from the swamp forests of Sundaland and adjacent Indochina look similar, recalling *Nannophlebia* especially with their swollen abdomen bases and arched tails, but such an affinity is unproven. Nonetheless, the pantalines' distribution suggests they arose on Gondwana (p. 111) before Africa and Australia split from Antarctica and drifted toward Asia (compare pp. 171 and 199).

Probably, the ability to secure their eggs was instrumental in this subfamily's evolution. *Zygonyx* attach them to rocks in the splash zone of waterfalls (p. 54) and *Elasmothemis* in stringlike filaments to rootlets. Leaftippers glue the eggs to leaves above the water, while in flight, and firebellies stick clumps to leaves hanging in the current. Males defend such spots vigorously, flashing the bright orange or yellow undersides and white-pruinose backs of their abdomens.

LIBELLULIDAE—DYTHEMISTINAE

SYLPHS, SETWINGS, AND KIN

BELOW | The abdominal club of this male White-tailed Sylph (*Macrothemis pseudimitans*) in Costa Rica may become covered entirely in white pruinosity.

With almost 140 species concentrated in the American tropics, this group seems like the New World's answer to Pantalinae (previous profiles): together, they make up about 35 percent of Libellulidae but over 70 percent of its flow-dependent species. Some *Dythemis* and *Brechmorhoga* adults are indeed uncannily like *Trithemis* and *Zygonyx*, the names setwing and dropwing referring to the same posture (p. 54).

No other libellulids, on the other hand, move quite as sylphs (*Macrothemis*) do. Contrary to what their scientific name suggests, they are rather small and slender, with males that skim rapidly over streams, landing close to the surface, usually flat on leaves, rocks, or sandbars. Away from water, they rest in an almost vertical position, their double-hooked claws recall those of other hanging (more distantly related) groups, notably *Macromia* (p. 80), hence their name.

With the similar little sylphs (*Gynothemis*), the larger and more consistently patrolling clubskimmers (*Brechmorhoga*), and the Antillean Clubskimmer (*Scapanea frontalis*), they form a group of over 60 tropical stream-dwellers.

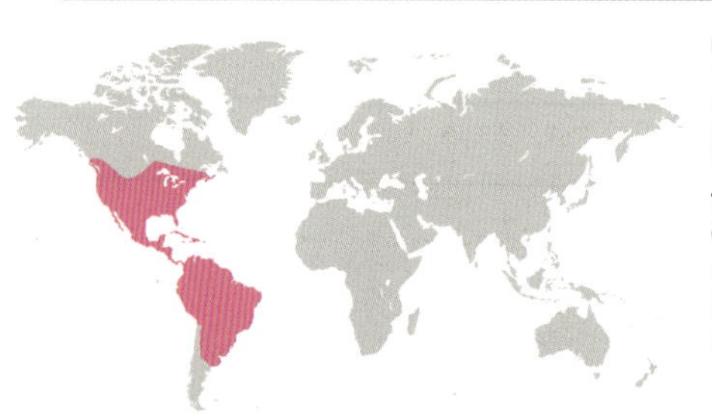

DIVERSITY
138 species in most standing and (especially) running waters in the New World's warmer reaches

TAXONOMY
Genera *Argyrothemis* (1 species), *Brechmorhoga* (16), *Dythemis* (7), *Edonis* (1), *Elga* (2), *Fylgia* (1), *Gynothemis* (4), *Macrothemis* (42), *Micrathyria* (48), *Nephepeltia* (6), *Pachydiplax* (1), *Paltothemis* (3), *Scapanea* (1), and *Zenithoptera* (4), but placement of *Nothodiplax* (1) uncertain

ABOVE | Checkered Setwing (*Dythemis fugax*) female in Texas, with its wings "set" to speed off.

BELOW | Red Rockskimmer (*Paltothemis lineatipes*) male in Arizona.

All feed over clearings and roads, multiple species often swarming together with females, which frequently have orange- or brown-marked wings, standing out.

As in many "stream-dragons" (pp. 74 and 92), the male's slender abdomen can be expanded near its tip, forming a club with a pale mark at its base. While both club and mark (typically located on the abdomen's seventh segment) are less frequent in libellulids than in other dragonflies, they are present in most dythemistines.

With black-checkered bodies infused with red (or dark blue pruinosity) and a tendency to perch vertically on rocks and walls, Middle America's rockskimmers (*Paltothemis*) recall *Bradinopyga* from Africa and Asia (p. 45). Their patrols over small streams and sustained feeding flights confirm they are affiliated to the sylphs, however.

While the subfamily forms the Neotropics' main libellulid radiation in running water, with *Erythrodiplax* their counterpart in more open stagnant sites there (p. 46), another nearly 60 dythemistines dominate well-vegetated or sheltered ponds and swamps. Most are recognized instantly by their glowing (blue-)green eyes, white face, and posture whereby they perch near the water with slender abdomen raised to show the tail-light and club, with wings drooped forward.

The species are mostly classified as Neotropical dashers (*Micrathyria*), but the Blue Dasher (*Pachydiplax longipennis*)—maybe North America's most numerous dragonfly—and minute dryads (*Nephepeltia*) belong here too. Brazil's Mantled Dasher (*Edonis helena*) and the barely known Canopy Skimmer (*Nothodiplax dendrophila*) from the Guianas may do too.

ABOVE | Piebald Sapphirewing (*Zenithoptera viola*) female in French Guyana. The blue-metallic reflections are startling to us but may actually provide camouflage against a backdrop of glistening water.

Other genera stand more on their own. The setwings (*Dythemis*) have a mixed habit, preferring streams but often breeding in standing waters too. Accordingly, they look like oversized dashers or, rather, like clubskimmers that perch with the abdomen up and wings down, thus "set" to take off again.

Named for Old Norse spirit guides *fylgja*, the Pearleye (*Fylgia amazonica*) is spritely indeed. Black with bright red tail and ghostly white eyes, the tiny males guard leaf-littered pools in the Amazon rainforest. The Silverspot (*Argyrothemis argentea*) lights up shady streams in these tropical lowlands with its white-pruinose thorax as do the fairies (*Elga*) with their brilliant blue eyes.

Sapphirewings (*Zenithoptera*), which breed in forest-rimmed marsh and ponds, are also called morpho dragonflies. While smaller than *Morpho* butterflies, they too seem entirely dark with the wings shut, startling the observer when the blue-metallic (often white-banded) upperside is revealed! Except for Australia's *Cordulephya* (p. 84), no other dragonflies perch with closed wings when mature, although sapphirewings often droop them (or just the front pair) down as they relax, sitting high up in the trees.

MEADOWHAWKS, FLUTTERERS, AND KIN

Odonates can react quickly to environmental change (p. 6). Few have amazed observers more, however, than the whitefaces (*Leucorrhinia*) from the cooler parts of Eurasia and North America (**dark pink on map**). Once the boreal realm warms up in spring, these small dragonflies will soon dash over still water, their bright snouts and wingtip veins standing out against their dark bodies.

In Europe, two species favor peat bogs and acidic fens, two like lakes, often with abundant vegetation such as lilypads, and a fifth has intermediate tastes, such as for lushly grown ditches. As humans transformed the landscape, all five suffered. By the late twentieth century, the lake species had been virtually wiped out in the west, while the bog-dwellers, though largely confined to protected areas, were still quite abundant.

Within the last quarter-century, however, fortunes have turned completely. Numbers of the bog species are collapsing, notably at their southern limit, while the lake ones have become locally common again, even at isolated sites. The fifth species has made gains in many places, but losses elsewhere.

As summers get hotter, bogs are drying out. Climate change, moreover, affects wind patterns, and the persistent easterlies in the spring of 2018

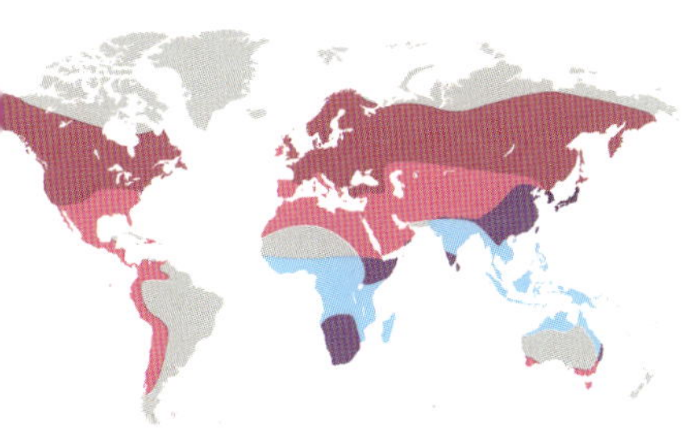

DIVERSITY
108 species found almost worldwide at mostly standing waters that, especially in cooler regions, are often comparatively warm

TAXONOMY
Genera *Austrothemis* (1 species), *Celithemis* (8), *Leucorrhinia* (14), *Nannophyopsis* (2), *Nesogonia* (1), *Rhyothemis* (23), *Sympetrum* (56), and *Tyriobapta* (3)

ABOVE | A male Lilypad Whiteface (*Leucorrhinia caudalis*) in Switzerland, uncharacteristically not perching on floating plants.

brought all five species to western locations where they had never been seen before. Water and landscape management also improved, restoring and creating habitats.

The advance of southern species (p. 46) must have mattered too, as may the interactions between the five relatives themselves and with their predators. The bog species' short-spined nymphs appear more vulnerable to fish than the long-spined lake species, for example.

Leucorrhinia is the most boreal genus of the family's largest temperate radiation, with about 80 species in three genera. While whitefaces merely have black marks at the wing bases, the closely related small pennants (*Celithemis*) of eastern North America's marshy ponds and lakes have colored faces and often elaborate wing patterns in brown, red, and gold, like painted glass. Resting atop vertical perches, they brandish these wings like pennants (pp. 65 and 67);

LEFT | Pair of Variegated Meadowhawks (*Sympetrum corruptum*) laying eggs in Montana, USA.

UPPER RIGHT | Emerald Bijou (*Nannophyopsis clara*) male in Hong Kong.

LEFT | The spectacular Picturewing Flutterer (*Rhyothemis variegata*) must be among South Asia's most photographed dragonflies.

LOWER RIGHT | North America's Halloween Pennant (*Celithemis eponina*) strongly recalls some of its distant tropical relatives, notably Australia's Graphic Flutterer (*Rhyothemis graphiptera*).

Celithemis translates as "stained-libellulid" (p. 38).

With 56 known species—called darters in the Old World and meadowhawks in the New—*Sympetrum* is the most successful odonate genus in the temperate north: their cold- and drought-resistant eggs let them benefit from the many standing or flowing microhabitats there that warm up quickly but can cool down or dry out just as fast. Females have very varied spout-like nozzles to place them, therefore, one Asian species even being called *S. cordulegaster* for its spiketail-like apparatus (compare p. 86).

They are good dispersers, some species migrating in great numbers. While most inhabit northeast Asia and North America, the genus has expanded south too, particularly in the Andes (**pink and purple on map p. 61, excepting Australia**). Eurasia's vagile Red-veined Darter (*S. fonscolombii*) extends across the higher and drier parts of Africa and has even established outposts on high islands such as Sri Lanka and Réunion.

Its large relative, *S. dilatatum*, occurred on the mid-Atlantic island of St. Helena, but was last seen in 1963, probably extirpated by introduced frogs. The Hawaiian Streamhawk (*Nesogonia blackburni*) has similarly adapted to island streams. While looking so distinct that it was placed in a separate genus, genetics show it originated from within *Sympetrum*. The evidence, moreover, suggests that this temperate radiation originally came from the Old World tropics.

RIGHT | Male of the Obsidian Flutterer (*Rhyothemis plutonia*) in Thailand. Translating to the inapt "stream-libellulid," the genus name may refer to rhyolite, a multicolored igneous rock.

Flutterers (*Rhyothemis*), formerly given their own subfamily Rhyothemistinae, are found at still waters with plenty of sun and vegetation (**blue and purple on map on p. 61**). They are the butterflies of the dragonfly world, with greatly expanded wings with extravagant dark markings. These, often interspersed with transparent windows and translucent amber, have strong blue, purple, and copper reflections, inspiring names such as Bronze, Iridescent, Obsidian, and Sapphire Flutterer.

Often abundant, these dragonflies clearly aim to be seen, sitting atop exposed perches, tilting their big wings from side to side in the breeze, like a funambulist on a tightrope, or weirdly angling their dark surface away from the sun. Aggregating to feed, they flutter about leisurely over clearings or shrubbery, often at some height. While seeming lackadaisical and weak, their flight can be very fast, and some species are strong migrants.

Treehuggers (*Tyriobapta*), by contrast, are restricted to Borneo, Sumatra, and the Malay Peninsula. They characteristically perch horizontally on tree trunks or rocks, camouflaged in blotched black and beige, like *Bradinopyga* (p. 45). Males become dark-glossy covered with bright pruinosity and fiercely defend small waterbodies in swamp forest. While two species have (almost) clear wings, the third has extensively dark hind wings with purple reflections.

The bijous (*Nannophyopsis*) of Southeast Asia have gleaming copper bodies and emerald eyes. Like the distantly related *Nannophya* (p. 49), from which their name derives, they are among the smallest dragons (down to ⅝ in/16 mm long). Perching deep inside weedy borders of ponds and lakes, with their yellow wings pressed forward and strongly clubbed abdomen curved down, they look almost like wasps.

Superficially, Australia's black-spotted red Swamp Flattail (*Austrothemis nigrescens*) is very distinct from its dark-metallic tropical cousins (**also pink on map p. 61**). Indeed, it is more like a temperate *Sympetrum*! The swampy pond and lake habitat, adult's paddle-shaped abdomen, and nymph's strange side-pointed eyes recall the bijous, however.

BASKERS AND KIN

Most insects cannot survive high concentrations of dissolved ions, but the closely related Black (*Selysiothemis nigra*) and Marl Pennants (*Macrodiplax balteata*), found along the Mediterranean and Caribbean coasts and in deserts nearby, fare well in brackish or alkaline waters. Their mature males' densely melanized skin (under a thin layer of reflective pruinosity) may help absorb intense heat and ultraviolet radiation, protecting the body beneath.

Like other so-called pennants (pp. 62 and 67), they sit atop exposed perches with legs thrust forward and wings raised, twisting in the breeze. The red Coastal Pennant (*M. cora*) breeds in lagoons in the Indo-Pacific, alongside 15 relatives known as baskers (*Urothemis*) and adjutants (*Aethriamanta*). These blue-pruinose or red species also inhabit warm and open sites, but favor richly vegetated standing freshwaters. The eggs of urothemistines can be bright green or blue and are often deposited on floating plants.

While *Urothemis* species are large and *Aethriamanta* small, this division may not hold up to closer scrutiny. *Selysiothemis* might best be sunk into *Macrodiplax*, thus losing a genus honoring the nineteenth-century founder of odonate taxonomy, the Belgian baron Edmond de Selys Longchamps.

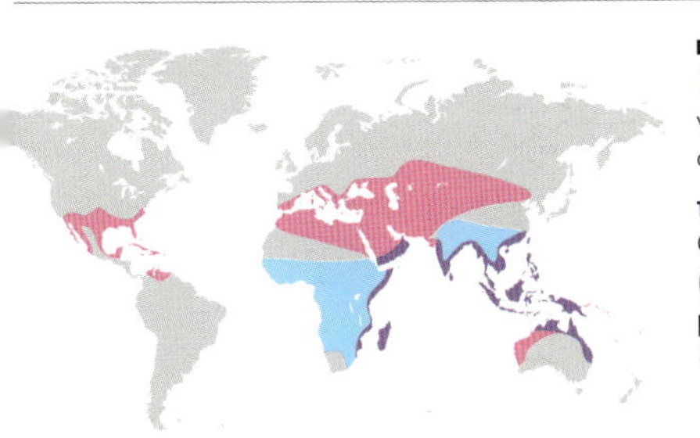

DIVERSITY AND OCCURRENCE
18 species of (mostly open) standing waters in warmer parts of the world, only one reaching the Americas

TAXONOMY
Closely related genera *Macrodiplax* (2 species) and *Selysiothemis* (1), **pink on map**; *Aethriamanta* (6) and *Urothemis* (9), **blue**

TOP | Coastal Pennant (*Macrodiplax cora*) male, perching at the end of a stick like a fluttering banner. Sometimes called Macrodiplactinae or even awarded family rank, the subfamily has distinctly wide-spaced venation.

ABOVE | Black Pennant (*Selysiothemis nigra*) male in Portugal.

AMBERWINGS, BLACKWINGS, AND POSSIBLE KIN

Entirely glossy black with deep red branding on the abdomen and sometimes face and wing bases, male blackwings (*Diastatops*) appear like glowing coals. They pursue each other in a rapid fluttery flight along the grassy verges of ponds and ditches in South America's tropical lowlands. At rest, the very broad and densely veined wings are often held at weird angles, possibly to prevent overheating, as males are known to abandon the waterside on hot afternoons.

If blackwings are smoldering embers, the tiny amberwings (*Perithemis*) are like flickering flames. The feisty males' wings and body appear entirely orange, run through with yellow to reddish veins and marked intricately with yellow and brown, like brooches of gold filigree. Female wings are more often clear, banded with amber and brown.

Like other compact libellulids with tinted wings, such as *Palpopleura* (p. 46) or *Zenithoptera* (p. 60), both sexes perch prominently and often high up, tilting one or both pairs of wings up or down, or waving them slowly. With a buzzy flight, they are easily mistaken for wasps or bees. All species favor smaller and often sheltered stagnant waters, like forest pools. Males typically perch in sunspots near the

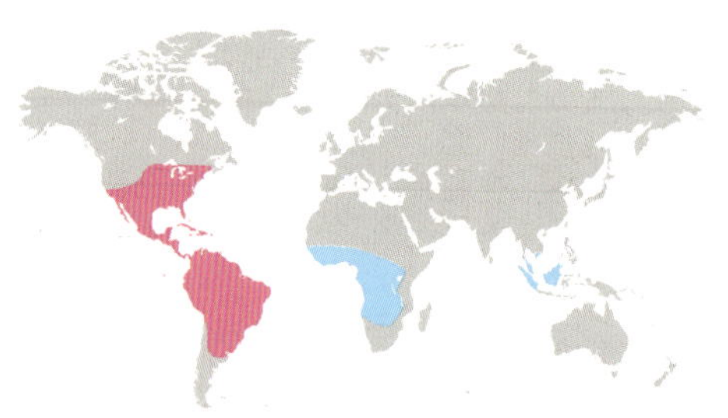

DIVERSITY
62 species of richly vegetated standing and slow-flowing waters, mostly in the American tropics

TAXONOMY
Genera *Brachymesia* (3 species), *Diastatops* (8), *Idiataphe* (4), *Oligoclada* (25), *Perithemis* (12), and *Planiplax* (5), but inclusion *Chalybeothemis* (3) and *Trithetrum* (2) speculative (**both blue on map**)

TOP LEFT | Red-saddled Blackwing (*Diastatops pullata*) male in Ecuador. Greek for "divided eyes," the genus name refers to the wide-spaced eyes, unique in Libellulidae.

LOWER LEFT | Mexican Amberwing (*Perithemis intensa*) male in Arizona, USA.

surface, taking off frequently to drive intruders away or inspect their little kingdom in a low hovering flight.

While both genera have slim legs, those of their apparent relatives of open water, the spiderlegs (*Planiplax*) and tropical (*Brachymesia*) and metallic (*Idiataphe*) pennants, are longer still. Bigger with less color in the wings, but often bright red bodies, all favor larger ponds and lakes. These pennants (compare pp. 62 and 65) rest at the tip of stems, raising their wings like flags. The restless males perch far out over water, readily speeding off for long skimming flights.

The small and slender leafsitters (*Oligoclada*), by contrast, spread their spindly legs across the top of leaves to oversee swampy backwaters and sluggish streams. Their dark bodies' slight steely sheen and light pruinosity creates a dull-metal look, contrasting with often gleaming green eyes. Found at weedy borders of lakes and rivers in Southeast Asia, greeneyes (*Chalybeothemis*) are remarkably alike. Equatorial Africa's Fiery (*Trithetrum navasi*) and Sooty Darters (*T. congoense*) are ecologically and structurally close, the former recalling a red *Brachymesia*. Further study, however, must confirm whether their long-mysterious affinities are truly Neotropical.

TOP RIGHT | Common Greeneye (*Chalybeothemis fluviatilis*) male in southern Cambodia.

LOWER RIGHT | A male Guiana Spiderleg (*Planiplax phoenicura*) showing off its long limbs in Brazil.

LIBELLULIDAE—TRAMEINAE

SADDLEBAG GLIDERS AND KIN

BELOW | As all species in the genus, this male Carolina Saddlebags (*Tramea carolina*) in Canada is a strong migrant. *Trameare*, indeed, is Latin for "passing through."

Flocks of Wandering Gliders (*Pantala flavescens*; p. 52) often contain similar dragonflies with wing markings flanking the body like bags on a saddle. Called gliders in the Old World too, but saddlebags in the New, the genus *Tramea* is indeed much like *Pantala*. The hind pair of the long, pointed wings is similarly broad and triangular, with notably smaller stigmas than the front. The big nymphs are spindly and thin-skinned, with bulging eyes and toothy protruding palps to quickly see and grab their prey, the abdomen ending in long and slender spines.

Closer scrutiny and genetics show they are not closely related. Both are adapted to seasonal rains, growing rapidly by feeding ferociously in balmy water, then setting off to find fresh opportunities. Adults are built to fly long distances with minimal effort, tending to glide and hover as they patrol their breeding sites or feed, before resting with a hanging posture. *Tramea* species favor more vegetated ponds and also often perch atop exposed lookouts with the abdomen held up or pressed down and wings raised to

DIVERSITY

39 species of mostly open standing waters in warmer parts of the world

TAXONOMY

Genera *Antidythemis* (2 species), *Atratothemis* (1), *Camacinia* (3), *Garrisonia* (1), *Hydrobasileus* (3), *Miathyria* (2), *Pseudotramea* (1), *Tauriphila* (5), and *Tramea* (21)

TOP RIGHT | The tricolored male of the Blazing Sultan (*Camacinia harterti*) seems to be on fire: rarely seen, this Asian behemoth probably breeds only between the buttresses of giant rainforest trees.

LOWER RIGHT | Hyacinth Glider (*Miathyria marcella*) male in Florida, USA.

show their markings. While *Pantala* females usually oviposit alone, *Tramea* males release and recapture their mate each time she swoops down to hit the water, unique behavior facilitated by their very long and slender claspers.

Although not all affinities are confirmed, aside from *Tramea* (**pink and purple on map**), another 18 conspicuous patrollers of standing waters, which readily wander and mostly rest hanging, are included in the subfamily too (**purple only**). Besides half of *Tramea* species, eight gliders in the closely related *Garrisonia*, *Miathyria*, and *Tauriphila* are Neotropical. They often mix with saddlebags in foraging flights, but are smaller and more colorful and favor floating vegetation (water lettuce and hyacinth, especially), their nymphs living among the roots. Amazonia's poorly known velvet gliders (*Antidythemis*) have unusually large stigmas.

The remaining diversity ranges from tropical Asia to the Pacific, although two Asian *Tramea* are common in Africa too. *Hydrobasileus* gliders and sultans (*Camacinia*) sail majestically over lush ponds and lakes on colored wings. Largely carmine and ebony, the massive Red (*C. gigantea*; photo p. 34) and Black (*C. othello*) Sultans appear like hefty hyperactive *Neurothemis* (p. 45). Hardly known, the Swarthy Sultan (*Atratothemis reelsi*) from Indochina and adjacent China could be allied, while *Pseudotramea prateri*, reported sparsely along the Himalaya, may just be a *Tramea* with smaller wing spots.

LIBELLULIDAE—ZYXOMMATINAE

DUSKDARTERS AND KIN

Visit any urban pond in tropical Africa or Asia by nightfall and you may see numerous dragonflies skimming fast and low over the foul water. Initially red with dusky-stained wings, eventually only dark dashes with whirring white flashes are visible in the gloom. Some without bright flashes tap the floating debris with their tail-ends, veer up, swing around, and tap down again, twisting to-and-fro as in a trance.

Named for the rhythmic motion with which its eggs are laid, the Twister (*Tholymis tillarga*) is among the tropics' most tolerant (even inhabiting brackish water) and widespread odonates; the Evening Skimmer (*T. citrina*) is its American counterpart. Daylight reveals a brown smear in each hind wing, paired in Twister males by a patch of gleaming pruinosity that looks ever brighter as darkness falls.

Flying continuously at dusk and (if warm enough) dawn, and hanging in vegetation all day, these are the quintessential crepuscular dragonflies, with big eyes, plain bodies, and broad

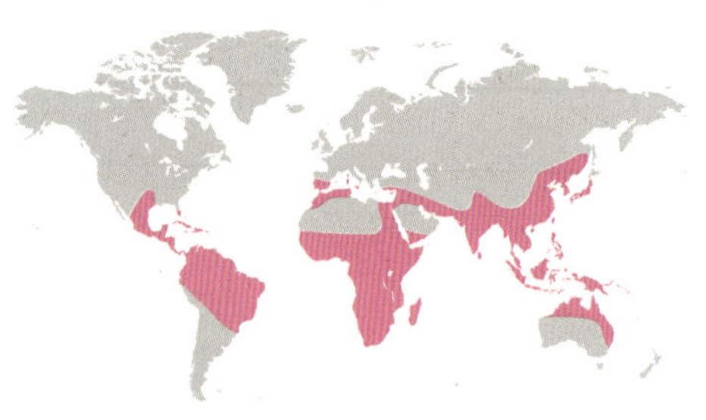

DIVERSITY
19 species of standing and slow-flowing waters in warmer parts of the world, but with only one reaching the Americas

TAXONOMY
Genera *Brachythemis* (6 species), *Deielia* (1), *Parazyxomma* (1), *Pseudothemis* (2), *Tholymis* (2), and *Zyxomma* (6), but placement of *Zygonychidium* (1) tentative

TOP | Following humans and other animals to snatch up disturbed insects, this Southern Banded Groundling (*Brachythemis leucosticta*) is among Africa's most familiar dragonflies.

LOWER | Male of the Twister (*Tholymis tillarga*) resting at daytime, showing the wing markings that stand out at dusk.

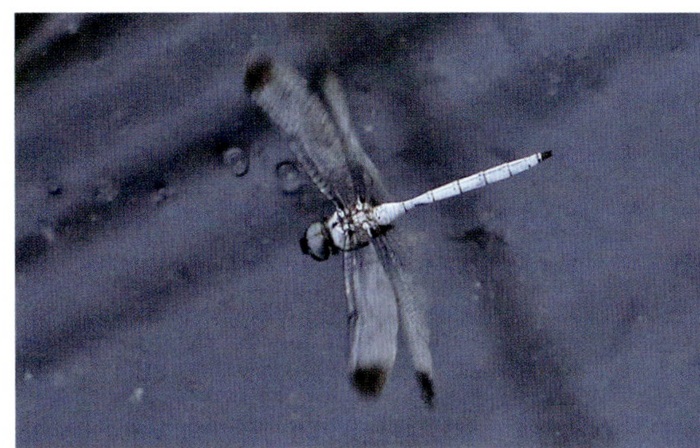

ABOVE | Whizzing over murky water at dusk, this Ghostly Duskdarter (*Zyxomma obtusum*) male on Sulawesi, Indonesia, stands out starkly due to the white pruinosity on its body and wings.

wings. Most closely related duskdarters (*Zyxomma*) from tropical Africa to the Pacific have fine-banded green eyes, like horseflies, and long narrow bodies, recalling small *Gynacantha* (p. 116).

The groundlings (*Brachythemis*) from the hotter parts of Africa and Eurasia, as well as *Deielia phaon* and the black-and-white cadets (*Pseudothemis*) from East Asia, look more normal, perching horizontally in full sun at (often bigger) still or slow-flowing waters, such as dams and rivers. They, especially the former, like bare habitats, settling on the ground or branches over water, often in great aggregations.

Although not specialized for twilight flight, many of these species become more active toward sunset and have barred eyes. Africa's Banded Duskdarter (*Parazyxomma flavicans*), furthermore, is structurally close to *Brachythemis* but behaves like *Zyxomma*! All genera have a rapid skimming flight over open water—often large expanses—and place their eggs in flight onto (plant) material lying on the surface. Their nymphs have big abdomens bearing distinct keel-like dorsal spines.

The bizarre Streamertail (*Zygonychidium gracile*), named for its ribbonlike claspers, is known from only two records on large rivers in western Africa. Various characters, including striped eyes, indicate it should be placed here too.

ABOVE | Surviving in even the filthiest water, Asia's Orange-winged Groundling (*Brachythemis contaminata*) is known as Ditch Jewel in India. Meaning impure, the species' scientific name seems appropriate too, but refers to its tinted wings.

RIGHT | Pied Cadet or Northern Bandtail (*Pseudothemis zonata*) male from Japan.

LIBELLULIDAE—TETRATHEMISTINAE

ELVES AND KIN

Once, close to 100 species in about 25 genera were placed here, but only 17 in three genera confined to the Old World tropics remain today. Rather than representing an ancestral group superseded by modern libellulids, as earlier experts believed, small dark species with slim wings and modified venation evolved multiple times (p. 42). The name *Tetrathemis* refers to a distinctive cell called the triangle that, within the limited space of its narrow wings, has four (*tetra* in Greek) sides.

Aside from Madagascar's Prettywing (*Calophlebia karschi*) and Africa's forestwatchers (*Notiothemis*), all species are currently classified as elves (*Tetrathemis*). The tiny black males perch above deeply shaded pools. Often only their gleaming blue eyes and some yellowish spots stand out in the gloom, but several species have broadly yellow wings. The Prettywing and Africa's Black-splashed Elf (*T. polleni*) even have largely dark wings. Entirely gray-pruinose when the male is mature, the latter is the only species that lives outside forest.

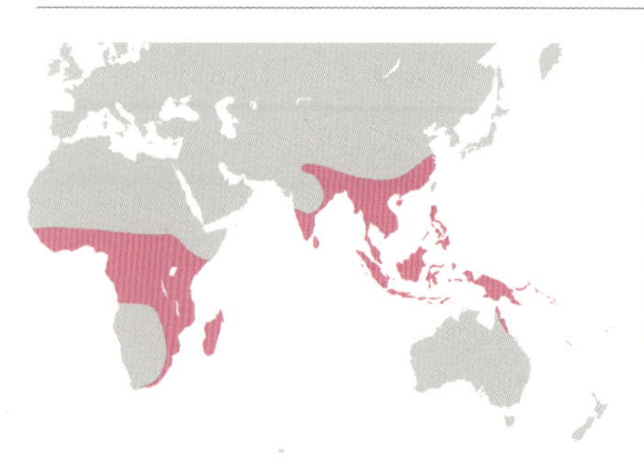

DIVERSITY
17 species of shady pools, mostly in forest, in the tropics from Africa to north Australia

TAXONOMY
Genera *Calophlebia* (1 species), *Notiothemis* (2), and *Tetrathemis* (14, but *Neophlebia* may be resurrected for distinctive African species)

TOP | The subfamily is unusual for attaching the eggs to material hanging above water, usually a twig. African *Tetrathemis*, such as this Black-splashed Elf (*T. polleni*), are among rather few libellulids that oviposit while settled.

LOWER | Like many odonates with marked wings, males of Madagascar's Prettywing (*Calophlebia karschi*) also occur in a clear-winged form.

LIBELLULIDAE—BRACHYDIPLACTINAE

PALEOTROPICAL DASHERS

This subfamily was once a grab-bag of over 130 smallish species in 25 genera, considered less primitive than Tetrathemistinae, but still lacking the heft and complex venation of presumably more advanced subfamilies (see opposite and p. 42). Only six other species seem close to the dashers or lieutenants (*Brachydiplax*), however.

These dashers are rather small nondescript dragonflies, with blue-gray pruinose males. They are common at sheltered, densely vegetated still waters in tropical Asia and the Pacific.

The pixies (*Brachygonia*) and imps (*Raphismia*) are tiny specialists confined largely to (peat) swamp forest on Borneo, Sumatra, and the Malay Peninsula. Tolerating somewhat saline water, only the Mangrove Imp (*R. bispina*) also extends to the Philippines and northeastern Australia.

The Inspector (*Chalcostephia flavifrons*) breeds in tropical Africa at most sunny swamps with tree-cover nearby, but is scarce on Madagascar (**blue on map**). Great numbers of adults often shelter in undergrowth, particularly in flood forest. They are rather boldly marked, but males become completely blue-pruinose, leaving only the diagnostic bright yellow face with green-metallic forehead.

While most libellulids have whitish to dull yellow or brown eggs, those of *Brachydiplax* and *Chalcostephia* are bright turquoise. Eggs are not preserved in collections and rarely photographed, so more knowledge remains to be gained.

ABOVE | Largely orange with a white-pruinose saddle at its abdomen base, this delightful Common Pixie (*Brachygonia oculata*) male from Malaysia guards small leafy pools.

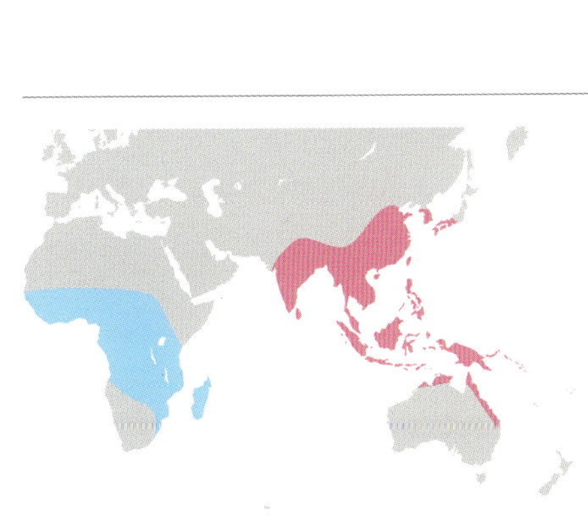

DIVERSITY
13 species of standing waters with rich vegetation or leaf-litter, often in or near forest, ranging from the tropics of Africa to the Pacific

TAXONOMY
Genera *Brachydiplax* (7 species), *Brachygonia* (3), *Chalcostephia* (1), and *Raphismia* (2)

FAMILY: CORDULIIDAE
EMERALDS

With brilliant green eyes and dark bodies often with a green-metallic sheen, most emeralds live up to their common name. Their scientific name can be translated as "clubbed-ones." In fact, all genera that William Leach named in 1815 for their expanded abdomen ends (*Cordulia*, *Cordulegaster*, and *Gomphus*) denote families today (pp. 86 and 92). This club is indeed frequent in dragonflies, particularly those that patrol when active and hang when settled (maybe it facilitates flight, perhaps as balancing weight) and that favor running waters.

Being more regular perchers, such species are rather scarce among Libellulidae: think of the clubskimmers in *Brechmorhoga* (p. 58) or *Zygonyx* (p. 54). About 440 such species, however, are close enough to be placed in the superfamily Libelluloidea, but retain too many features found also in more distantly related groups, such as Cordulegastridae and Gomphidae, to qualify for Libellulidae itself (p. 34). Auricles are another example: these earlike structures at the base of the male's abdomen may guide the female to his

LEFT | The scientific (meaning "bronze clubbed-one") and common names of Europe's Downy Emerald (*Cordulia aenea*) encapsulate many of the characteristics of the family Corduliidae.

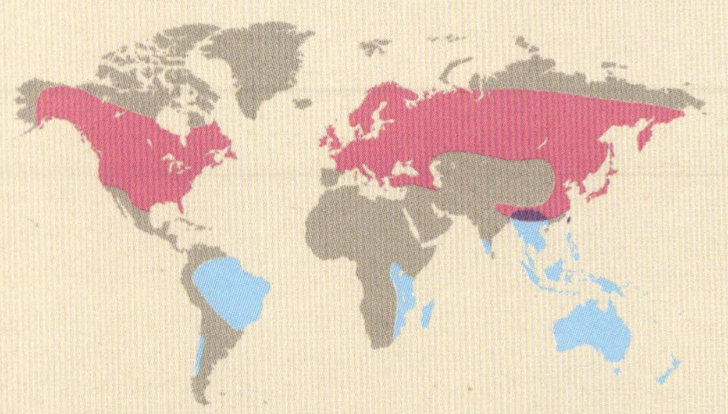

DIVERSITY
163 species in wide range of habitats worldwide, often cooler and/or more isolated than those where libellulids occur

ADULT HABIT
Mostly medium-sized dragonflies recalling libellulids, but typically fly continuously at the waterside; largely dark (often glossy) bodies and bright green eyes

TAXONOMY
Following this family's rearrangement, its subdivision remains unsettled: the well-defined group of northern genera (**pink on map**) is discussed separately from various groups found farther south (**blue**) for convenience

LEFT | The nearest dragonflies that look somewhat like this male Patagonian Emerald (*Rialla villosa*) from Argentina occur as far away as North America and New Zealand.

BELOW | A female Brilliant Emerald (*Somatochlora metallica*) from the UK with the fresh colors and shiny wings typical of newly emerged dragonflies, as well as the impressive egg-laying equipment of many corduliids.

genital opening. While present in the blueprint of all dragonflies, they have been lost from several well-dispersing groups of mostly standing water, such as *Anax* in Aeshnidae (p. 118) and all Libellulidae. Perhaps the sharp gap in the hind wing base needed to accommodate them impairs long-distance flight.

Although until recently these 440 species were thrown together as Corduliidae, they represent various distinct lineages that survive (mostly) where libellulids are less dominant, such as in isolated and cooler environments. These include sub-Arctic lowlands and mountain streams, but also continents like Australia and South America. Some relationships of these remnant groups remain uncertain: while the genera closest to Libellulidae are retained in Corduliidae, others have been placed in separate families (pp. 80–3) or are still taxonomically homeless (p. 84).

CORDULIIDAE—*CORDULIA* AND OTHER BOREAL GENERA

NORTHERN EMERALDS AND KIN

Over two-fifths of the 163 corduliids are fairly closely related and confined to cool-temperate parts of the Northern Hemisphere, particularly the deciduous and coniferous forest belts. Found in bog lakes above the Arctic Circle in Canada, Alaska, and Eurasia, where taiga gives way to tundra, the Treeline Emerald (*Somatochlora sahlbergi*) is the most polar odonate.

With dark and hairy bodies, adults can warm up rapidly in the sun and then retain their heat. During the day, many species fly in shadier places, however, their body's bronzy sheen perhaps shielding them from the brightest rays. Their eyes presumably have similar pigmentation, but they gleam more verdantly due to the light cast through them.

Found among living and dead plant material, the leggy nymphs vary almost as much as libellulids do, from small hairy critters hiding in dense peatmoss to big-spined beasts that brave

LEFT | Male of the Yellow-spotted Emerald (*Somatochlora flavomaculata*) in Germany.

TOP RIGHT | Unlike most corduliids, baskettails have extensively marked wings, none more so than the Prince Baskettail (*Epitheca princeps*), as this female from the USA shows.

LOWER RIGHT | Found only locally in eastern North America and early in the season, this male Ringed Boghaunter (*Williamsonia lintneri*) will mostly rest flat or upright, rather than hanging.

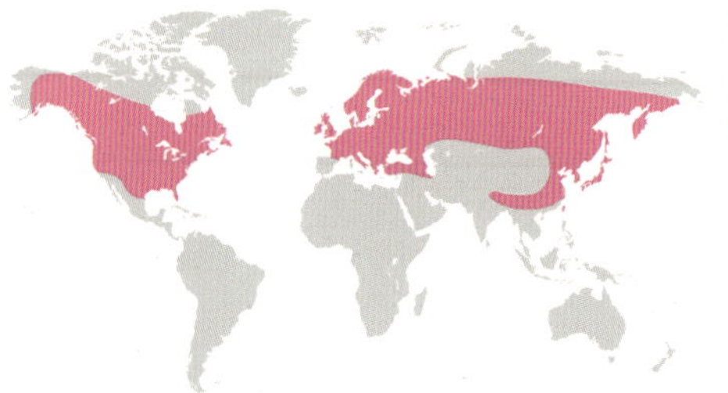

DIVERSITY
70 species in wide range of habitats in temperate zone of Northern Hemisphere

TAXONOMY
Genera *Cordulia* (2 species), *Corduliochlora* (1), *Dorocordulia* (2), *Epitheca* (12), *Helocordulia* (2), *Neurocordulia* (7), *Somatochlora* (42), and *Williamsonia* (2)

fish-infested lakes. Up north, only very rapid waters are avoided, but southward the species become restricted to streams in lowlands, and bogs and tarns higher up, probably because these are cooler with fewer libellulids.

Many females have impressive spike- or scoop-like lips below the abdomen tip to place eggs in wet moss or mud (see photo p. 75). Baskettails (*Epitheca*) push them out onto the large two-pronged plate for which they are named (the scientific name comes from the Greek for "cover" or "lid"). They seek suitable vegetation while carrying this egg mass below their upcurved tail-end. When a good place is found it is stricken off in flight, unraveling into a gelatinous strand that may swell to be an inch wide (2.5 cm) and several feet (over 1 m) long.

The group is richest in (particularly northeastern) North America, with 29 of the 47 typical emeralds (*Cordulia*, *Dorocordulia*, *Somatochlora*) and 10 of 12 baskettails present (the others inhabit Eurasia), as well as several unique genera. The tiny boghaunters (*Williamsonia*) are unusual for perching on the ground or trunks, rather like libellulids, while sundragons (*Helocordulia*) are limited to wooded streams and rivers. Shadowdragons (*Neurocordulia*) inhabit running waters (and large lakes) too, but are the temperate zone's only truly crepuscular odonates. Just like tropical twilight-fliers (pp. 70 and 116) they are large-eyed, broad-winged, rather uniformly brown, and active from dusk, flying about erratically into the night.

CORDULIIDAE—*HEMICORDULIA* AND OTHER AUSTRAL AND TROPICAL GENERA

SOUTHERN EMERALDS AND KIN

While northern corduliids could diversify across a wide swathe of circumpolar land (p. 76), habitat farther south is much scarcer and more fragmented. Most species there are placed in just two ill-defined genera, those in *Hemicordulia* recalling members of *Procordulia* but lacking auricles (p. 74).

Ecologically, they are indeed most like libellulids. Dispersing over great distances, the Australian (*H. australiae*) and Tau Emeralds (*H. tau*) occupy almost any open and standing water in Australia, while the Desert Emerald (*H. flava*) uniquely survives in its deepest interior. The group also extended across the Pacific and Indian Oceans,

LEFT | Mating wheel of the Yellow-spotted Swamp Emerald (*Procordulia grayi*) in New Zealand.

NEAR RIGHT | A male Fiji Emerald (*Procordulia irregularis*) living up to its name!

FAR RIGHT | Australia's Metallic Tigerhawk (*Pentathemis membranulata*) forms a relict group with its distant relatives in Madagascar and South America.

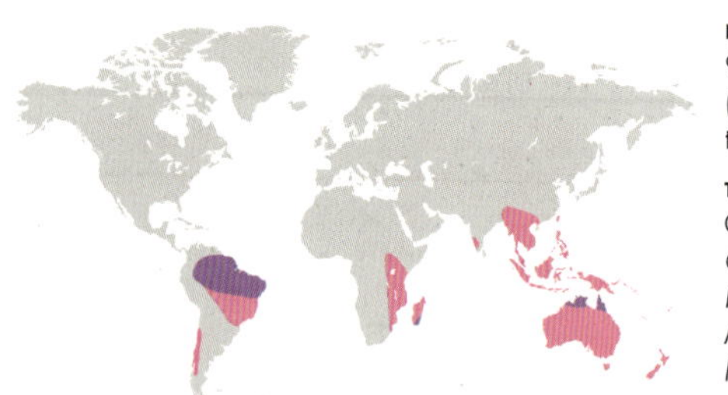

DIVERSITY

93 species of mostly standing waters (but also flowing), confined largely to the Southern Hemisphere

TAXONOMY

Genera *Antipodochlora* (1 species), *Cordulisantosia* (3), *Guadalca* (1), *Hemicordulia* (41), *Heteronaias* (1), *Metaphya* (3), *Navicordulia* (11), *Paracordulia* (2), *Procordulia* (18), and *Rialla* (1), **pink and purple on map**; more distinct *Aeschnosoma* (8), *Libellulosoma* (1), *Pentathemis* (1), and *Schizocordulia* (1) probably form an unnamed subfamily (**purple only**)

distinct species evolving on islands and isolated mountain ranges, even gaining a foothold in South Asia and East Africa. Where they meet libellulids, cooler or more marginal habitats are favored. Réunion's *H. atrovirens* only inhabits stream pools in deep shade or at great elevations, for example. In an unexplained case of "inverted insularity," the African Emerald (*H. africana*) does not breed in the Indian Ocean or great African lakes, and yet is never far from them.

Four localized genera found mostly in forests might represent an earlier wave of colonists across the Pacific: the Dusk Emerald (*Antipodochlora braueri*) on New Zealand, *Guadalca insularis* on the Solomon Islands, *Heteronaias heterodoxa* in the Philippines (adapted to very rapid water, rather like *Zygonyx*; p. 54), and three evaders (*Metaphya*) on Borneo, New Guinea, and New Caledonia. In South America, the Eluder (*Paracordulia sericea*) is limited to Amazonia, *Rialla villosa* to the Patagonian Andes (see photo p. 75), and *Cordulisantosia* species to the Atlantic Rainforest, while *Navicordulia* is scattered around Amazonia and bordering highlands.

The darnertails (*Aeschnosoma*, *Schizocordulia*) from Amazonia, Metallic Tigerhawk (*Pentathemis membranulata*) from northern Australia, and Skimmertail (*Libellulosoma minutum*) from Madagascar, finally, form a very distinctive but barely known group of corduliids. Until its rediscovery in 2016, for example, the latter was known to science for 110 years from just two specimens.

As their scientific names suggest, the adults' long black-and-yellow bodies may recall aeshnids or libellulids (pp. 34 and 112). The nymph's abdomen ends in exceptionally long lateral spines, perhaps serving as defense in their stream habitat. Perhaps these genera represent the last vestiges of a family that was once widespread in the tropics, possibly already becoming isolated in the early Cretaceous.

FAMILY: MACROMIIDAE
CRUISERS

Few species speak more to European dragonfly lovers' imagination than the Splendid Cruiser (*Macromia splendens*). Up to 3 in (7.5 cm) long and glossy black with emerald eyes and gold studding, adults are highly elusive, having usually sped by before you even realize it!

The species only survives locally on calm rivers in Iberia and southern France, probably since before the Ice Ages ravished Europe, but its nearest relatives occur as far away as North America and East Asia. Extending farther to New Guinea and northern Australia, some 80 species are placed in *Macromia* today. Two *Didymops* from eastern North America—ebony and ivory rather than jet and gold—probably fall within this vast, widely spread diversity too, which is in urgent need of taxonomic review.

All species are swift fliers. Males make long patrols over streams, rivers, and sometimes lakeshores. The furtive females dash out over the water, drop their eggs with a few quick taps on the surface, and then speed off into the trees again.

Only on the rare occasions that adults are found hanging in the vegetation can their numerous distinctive (but often technical, concerning wing venation) details be seen. The euphonious name *Macromia* refers to the great distance between

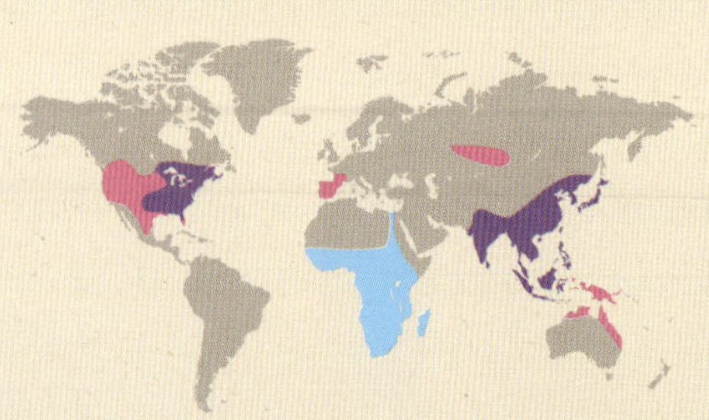

DIVERSITY
122 species of diverse running waters, but (larger) standing ones too; widespread in warmer parts of the world, but absent from American tropics

ADULT HABIT
Generally large dragonflies differing in venation details from most (former) corduliids (p. 74); often bigger and marked more boldly, with notably long legs, wings, and (clubbed) abdomen

TAXONOMY
Genera *Didymops* (2 species), *Epophthalmia* (5), **both purple on map**; *Macromia* (79), **purple and pink**, and *Phyllomacromia* (36), **blue**

RIGHT | Madagascar Cruiser (*Phyllomacromia trifasciata*) female showing the deep reddish club typical of many Malagasy odonates (p. 231) and the colored wings shared by many female dragonflies of forest streams (p. 91).

OPPOSITE | Splendid Cruiser (*Macromia splendens*) male in Portugal.

forewing base and node, translating from Greek as "long-shouldered."

The macromiids' very long legs stand out particularly on their nymphs' broad oval bodies (see photo p. 9). Although all species have this spidery look, as well as diagnostic protuberances between the antennae or behind the eyes, the nymphs are incredibly varied in size, body compression, markings, and the often fearsome dentition of the mask.

Each occupies a different microhabitat among the many created by varied rates of flow in stream habitats, ranging from hollowed-out banks to pool bottoms, and from fine silt to packs of coarse leaf-litter and dense mats of rootlets. As a consequence, multiple species may occur together at forested waters in Asia, and also in Africa, where 36 species of the ecologically similar *Phyllomacromia* are found.

The latter are related to Asia's pondcruisers (*Epophthalmia*). Rather like *Ictinogomphus* found at the same lakes and marshes (p. 102), these standing-water species are the largest members of a predominantly running-water family, the robust adults being up to 3⅓ in (8.5 cm) long. They must be top predators there too; the daggerlike teeth of the impressive nymphs are fanned out, like clawed hands from a horror movie.

FAMILY: SYNTHEMISTIDAE

TIGERTAILS, URFLIES, AND MISTFLIES

Of all former corduliids, tigertails are least like emeralds (p. 74). With small thoraces and long, narrow abdomens, they are weaker fliers and often hang on vegetation by their breeding habitats, which include a wide range of running waters, from boggy seeps to rocky streams and pools of intermittent rivers. Females may have notable structures—some rivaling those of cordulegastrids (p. 86)—for placing their eggs in such specialized haunts. The adults' eyes are frequently blue or brown. Their black or brown bodies are marked intricately with yellow, hence their common name.

Currently, 25 species appear confined to eastern Australia and adjacent Tasmania, and four to southwestern Australia. The genera *Archaeophya*, *Archaeosynthemis*, *Austrosynthemis*, *Choristhemis*, *Parasynthemis*, *Synthemiopsis*, and *Tonyosynthemis* are endemic, but *Eusynthemis* was also reported (but possibly misclassified) with one species from the Solomon Islands. *Calesynthemis* and *Neocaledosynthemis* are restricted to New Caledonia, while all but one species of *Palaeosynthemis* from the Moluccan island of Halmahera occur on New Guinea. Genetics and the disconnected distributions of two of these genera, however, suggest they need to be redefined.

LEFT | A male Yellow-tipped Tigertail (*Choristhemis flavoterminata*) from eastern Australia.

DIVERSITY
60 species at wide range of running waters from Tasmania to New Caledonia and the Moluccas and in the Andes of South America

ADULT HABIT
Medium-sized dragonflies recognized within range by combination (see main text) of color, build, wing veins, and habits, e.g. perch quite frequently at the water, but hang when they do so

TAXONOMY
Genera *Archaeosynthemis* (4 species), *Austrosynthemis* (1), *Calesynthemis* (7), *Choristhemis* (2), *Eusynthemis* (15), *Neocaledosynthemis* (2), *Palaeosynthemis* (13), *Parasynthemis* (1), *Synthemiopsis* (1), *Synthemis* (3), and *Tonyosynthemis* (2) form subfamily Synthemistinae if *Archaeophya* (2) and *Gomphomacromia* (5) placed in Gomphomacromiinae; *Pseudocordulia* (2) may represent further subfamily or separate family Pseudocorduliidae

The family's own limits are contentious too, moreover (p. 85). While in most Libelluloidea (see p. 74) the wing bases' central-most space is conspicuously devoid of cross-veins, synthemistids can be recognized by their presence. Eastern Australia's urflies (*Archaeophya*) lack these veins too and are therefore often placed in the subfamily Gomphomacromiinae with the similar American tigertails (*Gomphomacromia*). Confined to South America's Andes, the latter represent the only synthemistids outside Australasia. Nymphs live among wet moss or under moist rocks, sometimes several meters from open water in streams and seeps.

Unlike other Libelluloidea, nymphs of all forementioned genera grab prey with teeth lacking a fringe of hairlike bristles. Those of the mistflies (*Pseudocordulia*), which are almost terrestrial in leaf-litter along rainforest streams in northeastern Australia, have hairless teeth too. While the species (whose adults are wholly bronzy black) are occasionally placed in their own family, inclusion here may prove more appropriate.

TOP | Named for the male's round claspers, the Circle-tipped Mistfly (*Pseudocordulia circularis*) and its only close relative, the Ellipse-tipped Mistfly (*P. elliptica*) both from the tropical northeast of Australia, either belong with the synthemistids or to a family of their own.

LOWER | The "paradoxical" Patagonian Tigertail (*Gomphomacromia paradoxa*) vexed the first scientists who saw it, not being a gomphid-like macromiid, as its name suggests, but the only synthemistid genus beyond Australasia.

REMAINING GENERA FORMERLY IN CORDULIIDAE

MYSTICS, SHADESHIFTERS, AND ASSOCIATES

In the 1930s, zoologist Keppel Barnard encountered two probably ancient dragonflies at the tip of South Africa, botanically the most species-rich and disparate place on Earth. He called them *Presba* (*presbys* is Greek for "elder"), honoring the angler and entomologist who first collected them with the names *P. piscator* (Latin for "fisherman") and *P. venator* ("hunter"). The former was later revealed to be *Syncordulia gracilis*, described from an unknown locality a hundred years earlier.

Another century later, in 2006, this book's author found four species among the 13 specimens assembled by Elliot Pinhey, Africa's odonatological godfather. One was represented by just a single female, but luckily his colleague John Simaika had just caught another, mating with the unknown male! They christened the new finds *S. legator* ("gatherer"), recognizing their trailblazers, and *S. serendipator*, for all who are yet to discover serendipitously.

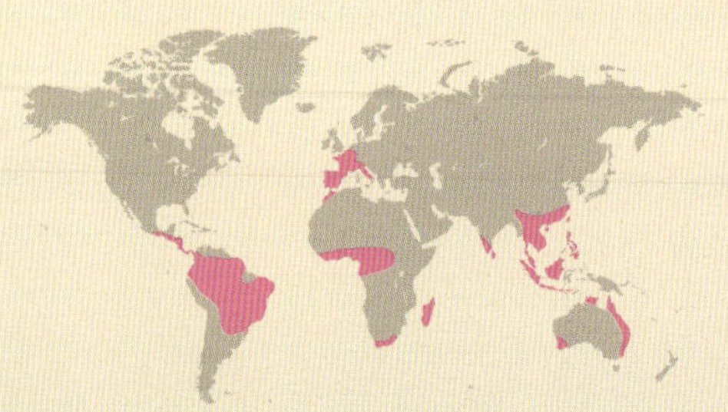

DIVERSITY
93 species of running waters found locally around the globe

ADULT HABIT
Mostly medium-sized dragonflies with look and habit of corduliids, but more varied in details of coloration and structure

TAXONOMY
Genera *Apocordulia* (1 species), *Austrocordulia* (3), *Austrophya* (2), *Cordulephya* (4), *Hesperocordulia* (1), *Idionyx* (29), *Idomacromia* (3), *Lathrocordulia* (2), *Lauromacromia* (6), *Macromidia* (10), *Micromidia* (3), *Neocordulia* (17), *Neophya* (1), *Nesocordulia* (6), *Oxygastra* (1), and *Syncordulia* (4) may constitute as many as 7 distinct families

These elusive presbas are typical of many dragons that became taxonomically homeless when Corduliidae was restricted (p. 74). The Orange-spotted Emerald (*Oxygastra curtisii*) was first found in Britain and named in 1834 for its great chronicler of insects, John Curtis, but has been extinct there for 60 years. Some experts award it a family of its own, Oxygastridae, but other orphaned genera may belong there too, such as *Syncordulia*. The dorsal keel of the male abdomen's tip is also present in tropical America's keeltails (*Neocordulia*) and Madagascar's knifetails (*Nesocordulia*), yet absent in equatorial Africa's shadowcruisers (*Idomacromia*), which are believed to be close.

As with other relict groups (pp. 110 and 122), some of the most distinctive unplaced species are Australian. Shutwings (*Cordulephya*) are the only dragonflies to consistently rest with folded wings. Confined to the dry interior's Murray-Darling Basin, the large-eyed Nighthawk (*Apocordulia macrops*) flies only in twilight. Ultimately, these genera may be grouped with their compatriots—the Rainforest Mystic (*Austrophya mystica*), Orange Streamcruiser (*Hesperocordulia berthoudi*), swiftwings (*Lathrocordulia*), hawks (*Austrocordulia*), and mosquitohawks (*Micromidia*)—in Synthemistidae (p. 82) or classified separately as Cordulephyidae and perhaps Austrocorduliidae.

Other genera may be closer to Macromiidae (p. 80) or deserve families of their own, such as tropical Africa's inch-long Feeblewing (*Neophya rutherfordi*), fluttering over rainforest streams on wide amber wings like a phantom butterfly. Shadowdancers (*Idionyx*) and shadeshifters (*Macromidia*) similarly flit furtively over dark forest runnels in tropical Asia. One *Lauromacromia* species occurs on South America's Guyana Shield, and five in southern Brazil. Most of the few specimens known were lost in 2018, when fire destroyed the national museum in Rio de Janeiro.

OPPOSITE | Adult shutwings, like this Common Shutwing (*Cordulephya pygmaea*) from Australia, only emerge toward the end of summer and in the fall. They are the only dragonflies to always rest with (nearly) closed wings, fore- and hind wings being almost identical in shape.

TOP RIGHT | Decorated exquisitely with ivory, the Mahogany Presba (*Syncordulia venator*) is found only at cool clear streams at Africa's extreme southern tip.

LOWER RIGHT | The male of *Idionyx corona* is one of many shadowdancer species restricted to southern India.

FAMILY: CORDULEGASTRIDAE
SPIKETAILS AND GOLDENRINGS

BELOW | Nymph of the Italian Goldenring (*Cordulegaster trinacriae*).

All profiles so far cover just one superfamily, Libelluloidea (p. 74), which encompasses nearly a quarter of odonate diversity, with almost 1,500 species. What might explain their success? While other nymphs have flat masks, theirs are scoop-shaped with broad triangular palps for spooning up many small prey at once. As in the gomphids, the second-largest dragonfly radiation (p. 92), the ovipositor is reduced (p. 110), allowing for the diverse ways of egg-laying presented on the previous pages (p. 18). Both features are shared with this family and the two treated next, forming the superfamily Cordulegastroidea, and so the two superfamilies are collectively called Cavilabiata, Latin for "hollow-lipped." The smaller superfamily's 105 species are restricted to running and mostly cooler waters, Cordulegastridae even being the most northerly distributed odonate family, extending south only in the mountains of Central America, North Africa, and South Asia. This localized superfamily has some

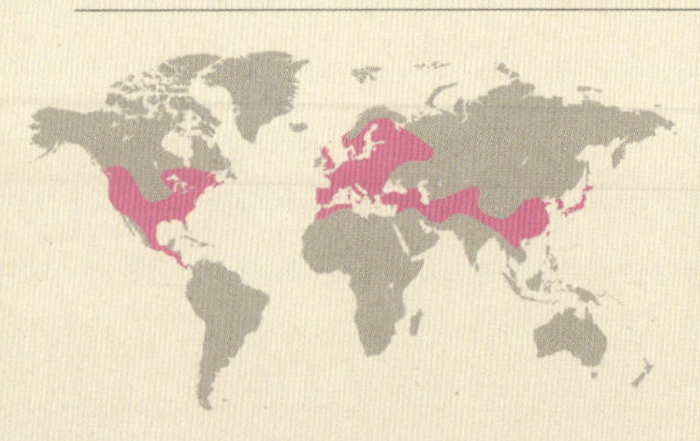

DIVERSITY
49 species of often smaller running waters in cooler reaches of the Northern Hemisphere

ADULT HABIT
Large dragonflies recognized easily within most of range by great size, long and dark yellow-spotted bodies, contrasting green or blue eyes that barely touch on top, and especially the egg-laying spike at female's abdomen tip

TAXONOMY
Genera *Anotogaster* (11 species), *Cordulegaster* (10), *Neallogaster* (8), *Thecagaster* (11), and *Zoraena* (9)

of the order's oddest egg-laying structures and behaviors. Known as goldenrings in Eurasia, in America spiketails were named for the diagnostic pointed plate at the female's genital opening. This projects well past the abdomen tip and is used in their unique egg-laying behavior (see photo below).

Searching for females, males fly long, often slow, beats low over the water. Like the egg-laying females, they are very focused and easily observed. Both sexes may perch near water close to the ground, flying up after prey, or hunt in rapid flight along forest roads.

The thick but long-bodied nymphs (see photo p. 32) are instantly identifiable within most of their range, having very jagged and unequal teeth on their palps. Being sluggish and retiring, they seek out the slowest-moving water and bury themselves in the soft substrate, their hairy bodies typically covered in muck.

Ranging from mossy trickles to (smaller) rocky rivers, habitats are often shaded and occasionally intermittent. Found even on barren streams high in the Himalayas, they must owe their boreal success to their cold tolerance, perhaps taking three to seven years to develop. Indeed, they are often the only dragonflies in their habitats.

The species look very similar and while most east and south Asian ones are placed in *Anotogaster* and *Neallogaster*, most others were long lumped in *Cordulegaster*. Most American species were recently moved to *Zoraena*, however, and many Eurasian ones to *Thecagaster*.

TOP RIGHT | This male Giant Goldenring (*Anotogaster sieboldii*) from Japan is one of the largest dragonflies. The cordulegastrids' long, clubbed abdomen (*Cordulegaster* translates from Greek to "club-belly") always droops at rest.

LOWER RIGHT | Hovering over a shallow spring, this female Blue-eyed Goldenring (*Thecagaster insignis*) from Turkey may bob up and down 500 times or more, jabbing her eggs into the bottom one by one.

FAMILY: NEOPETALIIDAE

FUNNELTAILS

DIVERSITY
Single species of seeps and streamlets in the temperate forests of Chile and adjacent Argentina

ADULT HABIT
Rather large dragonflies with black bodies with double rows of greenish spots, narrowly touching eyes, orange stigmas and four red blotches on the leading edges of each wing, peculiar black tufts on the undersides of the abdomen segments, and a huge funnellike expansion of the female's abdomen tip

TAXONOMY
Genus *Neopetalia* (1 species)

When first describing the Funneltail (*Neopetalia punctata*) from Patagonia in 1854, Selys (p. 65) placed it near Cordulegastridae (p. 86). Discovery of the similarly archaic and spot-winged austropetaliids there, however, led him and subsequent authors to consider the species closer to those (p. 128). Only when the nymph was found in the 1990s was his original hunch confirmed!

Despite 3,000 miles (5,000 km) between them, the nymphs are indeed surprisingly alike, including their irregular teeth, probably three-year development time, and the habit of burying themselves into the silty substrate of seeps and streamlets. The Valdivian Forests of Chile and adjacent Argentina in which they occur are the austral equivalent of the temperate landscapes that cordulegastrids inhabit, of course.

Although adults are superficially similar, they differ from cordulegastrids in many respects. The face is especially wide, and the hairy body is broader and shorter, as are the red-blotched wings. Adults also behave differently, being most active when the sky is overcast, flying high (between 15 and 50 ft/5 and 15 m) above the breeding habitat.

Their oddest feature, however, is the almost circular and rather membranous "funnel" at the female's abdomen tip that, supported by two claw-like terminal appendages, must aid oviposition. Perhaps this "splash plate" allows for a large egg mass to be built up and then spread quickly. Indeed, females have been seen hovering briefly above the water, then smacking it with their tail-end. That too, of course, cannot differ more from cordulegastrids, which sow their eggs laboriously.

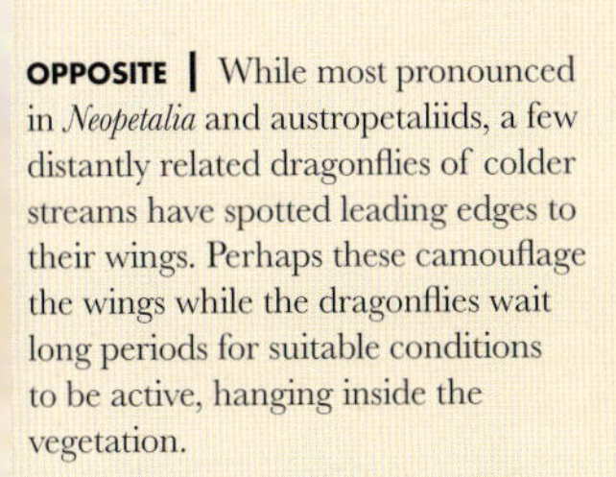

OPPOSITE | While most pronounced in *Neopetalia* and austropetaliids, a few distantly related dragonflies of colder streams have spotted leading edges to their wings. Perhaps these camouflage the wings while the dragonflies wait long periods for suitable conditions to be active, hanging inside the vegetation.

RIGHT | Patagonia's Funneltail (*Neopetalia punctata*) may be the world's most disparate dragonfly.

FAMILY: CHLOROGOMPHIDAE

SKYDRAGONS

LEFT | Female of the Butterfly Skydragon (*Chlorogomphus papilio*) from southern China.

BELOW RIGHT | The female of *Chlorogomphus aritai* from northern Vietnam.

When speaking of dragonflies' color and charisma we are often referring to males, dazzling each other and their mates (and us) with their looks (p. 13). While chlorogomphid males are impressive, however, the females steal the show, stunningly marked on often greatly expanded wings. Gliding at considerable heights while feeding, they appear like gigantic butterflies, with wingspans up to 5 in (13 cm). Some have wings infused with caramel, or patterned boldly with bars of chocolate and dabs of cream; other species wear a charcoal mantle trimmed with an ashy fringe. Many also have clear-winged forms of the female, while males of the Butterfly Skydragon (*Chlorogomphus papilio*) have piebald wings too.

Many species are named for their beautiful, golden wings (*C. auratus*, *C. auripennis*, *C. caloptera*, *C. xanthoptera*) or to express awe (*C. magnificus*, *C. preciosus*, *C. speciosus*, *C. splendidus*). But why be so showy? Typically, odonate females try to avoid

DIVERSITY
55 species of often forested streams and rivers in tropical Asia

ADULT HABIT
Large dragonflies resembling Cordulegastridae with their long black bodies marked with yellow bands and rings, but green eyes distinctly (although slightly) separated on top and all spaces in wing bases with some cross-veins; females lack the egg-laying spike and often have stunningly marked expanded wings

TAXONOMY
Genera *Chlorogomphus* (47 species), *Chloropetalia* (4), and *Watanabeopetalia* (4); more proposed genera may be valid

attention from predators or harassing males, thus being inconspicuous. Males, by contrast, must advertise themselves or their territories to potential mates, or their possession of such sites to rival males.

By slowly but incessantly patrolling streams, chlorogomphid males are looking for action rather than trying to attract it. Like many stream-trailing dragonflies, they don't need to stand out as much, therefore. Females might have to, though, particularly if male densities are low. If many similar species are around, furthermore, they may also benefit from being easy to identify.

Indeed, many other tropical dragonflies with clear-winged males, especially of forest streams, also have colorful females. Most *Macrothemis* (p. 58), *Anotogaster* (p. 86), and *Oligoaeschna* (p. 127) females have wings variably adorned with amber and brown. The biggest *Zygonyx* females soar above the forest on enlarged, distinctly patterned wings, just as skydragons do (p. 54).

Probably, female showiness evolved to such extremes in chlorogomphids because they are so big (larger animals occur in lower numbers) and diverse, with most species packed into a relatively small area in southern China and adjacent Indochina. Most are scarce, inhabiting highland forest runnels and streams, and thus vulnerable to human impacts. Females, moreover, are sought by collectors. Other species of lowland rivers appear more tolerant and numerous, especially in spring.

Lacking the related cordulegastrids' egg-laying spike (p. 86), females oviposit in shaded shallows while settled or fluttering about nervously. With long bodies and jagged teeth, the nymphs recall their relatives and presumably have similarly long development times. A unique shelf between the antennae may help them shovel through the gravel and sand in which they hide, emerging to hunt toward dawn and dusk.

FAMILY: GOMPHIDAE
CLUBTAILS

LEFT | Largely extinct by the 1980s due to heavy industrialization, the River Clubtail (*Stylurus flavipes*)—seen here having just emerged from its nymphal skin—recovered from the 1990s onward all across Europe in response to the clean water laws and was therefore included in IUCN's novel Green List of species (p. 30).

TOP RIGHT | A female of North America's Zebra Clubtail (*Stylurus scudderi*) pressing out a batch of eggs, which she will drop quickly at a suitable spot on the river surface.

When the River Clubtail (*Stylurus flavipes*) recovered all across Europe due to clean water laws, that not only reflected the power of Odonata to represent ecological integrity (p. 4), but also the unique character of Gomphidae: while "pond-dragons" (p. 34) and "pond-damsels" (p. 132) could diversify particularly due to their dominance in standing waters, these "stream-dragons" were the only large family to radiate almost exclusively in running ones.

The key was their ability to adapt to all substrates that moving water creates, reflected in an unparalleled diversity of body shapes. Some nymphs are stocky and crouch behind pebbles. Others are so flat that they look just like the leaf-litter they hide among. Most, however, bury themselves. Many, including *S. flavipes*, have strong legs to dig into soft river bottoms, for example, their drawn-out abdomens protruding above the sand or mud to take in clean water (see photo p. 9).

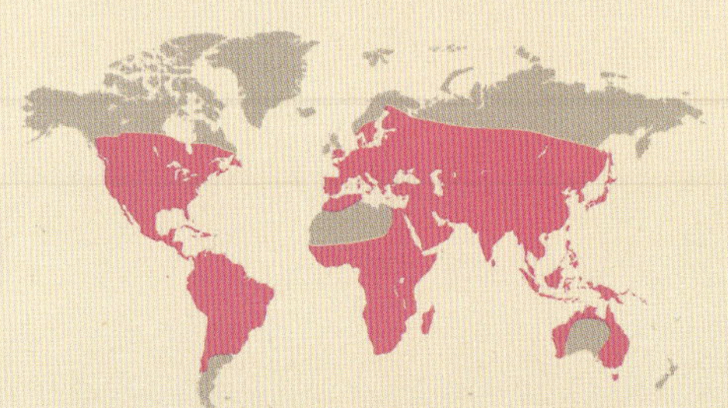

DIVERSITY
1,008 species of mostly running waters worldwide

ADULT HABIT
Small to very large dragonflies, typically with an expanded abdomen tip that is often accentuated with leaflike flaps or flanges called foliations; they are the only dragons with widely separated eyes in most parts of the world

TAXONOMY
Many subfamilies and tribes have been recognized but not all are well-defined. Most species may belong to just 4 large subfamilies (pp. 94–105), while 7 smaller groups discussed more briefly might each represent a subfamily too (pp. 106–09)

With their distinct layout, the antennae are particularly versatile: while in other families both adults and nymphs typically have threadlike antennae with many similar sections, those of gomphid nymphs have only four segments, the third much longer and even dilated in some. This variety likely serves distinct sensory functions in the many substrates inhabited—for example, to feel vibrations or smell.

While most stream odonates carefully place their eggs inside their preferred substrate, gomphids typically drop batches into the water near the nymphs' microhabitat. A sticky exterior or threadlike anchor may then ensure that they do not drift too far. This habit, shared with many libellulids, facilitates wider access to the many habitats within river systems, including those far from the banks, but also allows females to limit their time at the water.

Provided with few strategic options to intercept females, however, gomphid males are rather unshowy, both sexes often appearing and behaving rather alike (but see p. 98). Many species spend most of their time in fields, clearings, or the canopy, coming to water for only a few sunny moments, or just toward dusk, and are thus much harder to find as adults than as nymphs. Each encounter may be special, therefore, especially in the species-rich tropics, so gomphids are the favorites of many dragonfly experts.

ABOVE | Gomphids can be recognized instantly by their wide-spaced eyes (p. 99), usually dark, yellow or green marked body, and often clubbed abdomen. The latter may bear impressive foliations, as in this male Helmeted Leaftail (*Phyllogomphus coloratus*) from Gabon.

GOMPHIDAE—GOMPHINAE

TYPICAL CLUBTAILS

While their nymphs are especially varied, gomphid adults seem surprisingly alike (p. 92). The differences are simply more subtle, though, and the nymphs' proportions are often expressed in the adults. In tropical Africa's siphontails (*Neurogomphus*) and snorkeltails (*Mastigogomphus*), for example, the nymphs are stretched most extremely to facilitate burrowing deep into river bottoms, with stumpy legs and the abdomen's last segment shaped like a drinking-straw. Adults are short-legged too, with unwieldy abdomens. They cannot easily make sallies from a perch and hang clumsily in vegetation, probably feeding in flight over water instead. For similar reasons, Eurasia and North America's deepest burrowers, *Stylurus flavipes* and its relatives (see photos pp. 92–3), are known as hanging clubtails.

Owing to their superficial likeness, *S. flavipes* as well as over half of the 88 species making up the tribe Gomphini, were once grouped under *Gomphus*. The name derives from the Greek word for "bolt," as used in building, referring to the club-shaped abdomen (compare p. 74). The 38 in North America proved genetically closer to their compatriots, the pond clubtails (*Arigomphus*) and spinylegs (*Dromogomphus*) and were moved to

LEFT | This male of Europe's Yellow Clubtail (*Gomphus simillimus*) represents the archetypical gomphid, black-and-yellow with a clubbed abdomen: indeed its scientific name means "much alike!"

OPPOSITE TOP | Male of the Thai Bowtail (*Macrogomphus kerri*) in central Vietnam. The greatly extended penultimate segment of the abdomen is shared with the nymph, where it serves as a breathing-tube.

OPPOSITE BOTTOM | Left by the emerging adult along Africa's great Congo River, the nymph's skin demonstrates the stretched snorkel-like body of the Black Siphontail (*Neurogomphus martininus*).

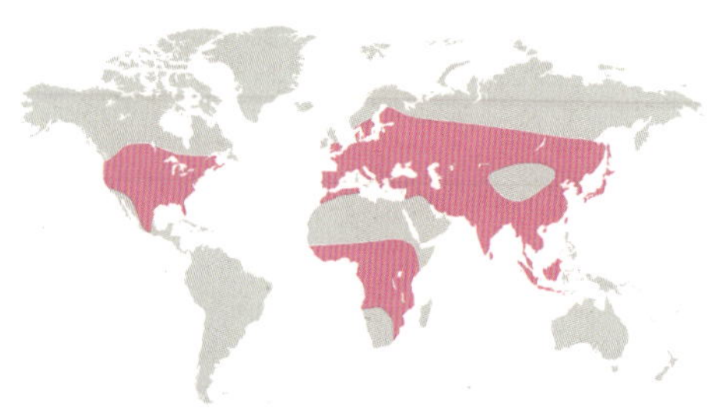

DIVERSITY
228 species of streams, rivers, and larger standing waters in North America, Africa, and Eurasia

TAXONOMY
Tribe Anisogomphini with genera *Anisogomphus* (15 species), *Euthygomphus* (9), *Labrogomphus* (1), *Mattigomphus* (2), *Merogomphus* (7), and *Notogomphus* (20); Cyclogomphini with *Anormogomphus* (3), *Asahinagomphus* (1), *Burmagomphus* (30), *Cyclogomphus* (5), *Platygomphus* (3), and possibly *Stylurus* (27); Gomphini with *Arigomphus* (7), *Asiagomphus* (28), *Dromogomphus* (3), *Gastrogomphus* (1), *Gomphurus* (13), *Gomphus* (8), *Hylogomphus* (6), *Phanogomphus* (17), *Shaogomphus* (3), and *Stenogomphurus* (2); Macrogomphini with *Macrogomphus* (15); and Neurogomphini with *Mastigogomphus* (3) and *Neurogomphus* (14)

Gomphurus, *Hylogomphus*, *Phanogomphus*, and *Stenogomphurus*, while *Gomphus* was restricted to just eight species from Europe and adjacent Africa and Asia. This tribe is the foremost in streams and rivers of the Northern Hemisphere's temperate zone, the fairly flat nymphs burrowing shallowly into fine sediments with much organic material, in ponds and lakes too. Aside from eastern North America, they are represented best in southern and eastern Asia, with 32 species placed in *Asiagomphus*, *Gastrogomphus*, and *Shaogomphus*.

The other tribes are richest there too. Adults of all 53 Anisogomphini have very long hind legs adorned with fierce spines, probably to catch large prey. African longlegs (*Notogomphus*) occur at Africa's higher elevations; the Asian longlegs (*Anisogomphus* and *Mattigomphus*) and spinelegs (*Euthygomphus* and *Merogomphus*) are confined to Asia.

South Asia's ferocious Marauder (*Labrogomphus torvus*) has similar legs, but also the extended second-last abdomen segment (in nymphs and adults) characteristic of the bowtails (*Macrogomphus*). Those are tropical Asia's most notable deep-diggers, placed in their own tribe, Macrogomphini. As North America's *Dromogomphus* are long-legged too, but clearly belong to Gomphini (see above), even the most conspicuous characters may say little about relationships.

Disentangling them and reclassifying the many species will take time, therefore. Even the allies of *Stylurus* are still uncertain. While the tribe of Africa's deepest-diggers, Neurogomphini, is not ruled out yet, recent studies propose Cyclogomphini, limited otherwise to the warmer parts of Asia: clenchtails (*Anormogomphus*) and picktails (*Platygomphus*) favor large rivers, splaytails (*Burmagomphus*) and sickletails (*Cyclogomphus*) mainly forest streams, while *Asahinagomphus* is hardly known.

GOMPHIDAE—HAGENIINAE

DRAGONHUNTERS, LYRETAILS, AND ALLIES

Although many dragonflies occasionally slay another dragon, few specialize in big prey. Some libellulids do prefer butterflies and dragonflies (pp. 36 and 49), though, as do the largest gomphids. The Dragonhunter (*Hagenius brevistylus*) has even been seen taking a hummingbird!

Up to 3½ in (9 cm) long, adults look like bodybuilders, their heads dwarfed by muscular bodies. With their bulk and long legs, both sexes perch awkwardly, preferring long and leisurely patrols along rivers and lakeshores in search of mates and prey. Hiding among detritus, the flattened nymphs are just as strange: with their almost circular abdomens and antennae, they resemble chips of wood.

This species and those of East Asia's equally impressive *Sieboldius* were long seen as the only Hageniinae (**purple and pink on map**). Genetics showed that these giants fall among some of the smallest gomphids, however, which were once

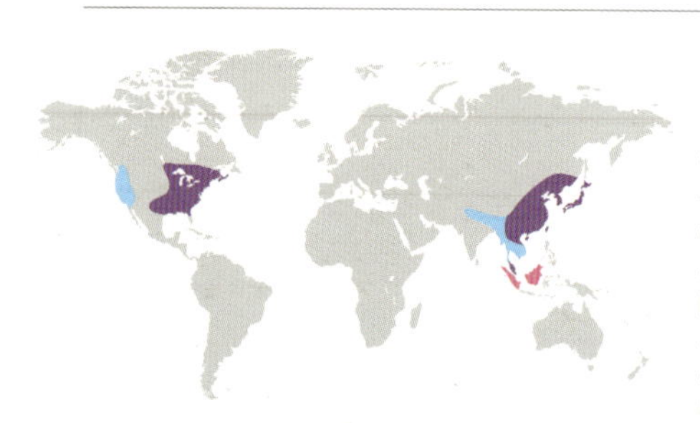

DIVERSITY
81 species of streams, rivers, and lakes in North America and especially South and East Asia

TAXONOMY
Genera *Hagenius* (1 species) and *Sieboldius* (8) are typical of subfamily, but *Davidius* (22), *Dubitogomphus* (3), *Lanthus* (3), *Octogomphus* (1), *Sinogomphus* (11), and *Stylogomphus* (15), often placed in Octogomphinae, appear close; Octogomphinae's *Fukienogomphus* (3) and *Trigomphus* (14) might be near Gomphinae (p. 94)

OPPOSITE | Contrary to its scientific name, the Sunda Dragonhunter (*Sieboldius japponicus*) occurs in Southeast Asia, this male photographed in Malaysia.

ABOVE | A female American Dragonhunter (*Hagenius brevistylus*) perching awkwardly to devour a smaller clubtail species.

BELOW | This male Patrolling Lyretail (*Stylogomphus shirozui*) from Taiwan is typical of the many smaller species in this group.

placed in the subfamily Octogomphinae (**purple and blue**). Found also in sand and gravel, their flat larvae with oval antennae are quite similar.

The Grappletail (*Octogomphus specularis*) is confined to North America's west; two species each of the pygmy clubtails (*Lanthus*) and least clubtails (*Stylogomphus*) occur in the east. The latter two genera (there called lyretails), with the clingtails (*Davidius*) and poorly known *Dubitogomphus* and *Sinogomphus*, encompass about 50 species in South and East Asia.

While the toothtails (*Trigomphus*) and tusktails (*Fukienogomphus*), also found there, were affiliated with Octogomphinae as well, they share similarities with the co-occurring *Macrogomphus* in Gomphinae (p. 95). Diversifying especially in the Northern Hemisphere's cooler parts, these groups are clearly close but their relationships must be studied further.

HOOKTAILS AND KIN

BELOW | Unusually well-adapted to temporary water for a gomphid, the Common Hooktail (*Paragomphus genei*) occurs across Africa and southern Europe. The nymphs develop within a few months in pools left by intermittently flowing streams and rivers.

While we quickly notice various dragonflies at a pond, often only damselflies are at first apparent at streams or rivers: many dragons only fly by (p. 74) or visit very briefly. Although the latter is true for many gomphids (p. 92), others are more like libellulids, with males perching near good egg-laying spots to intercept approaching females.

Hooktails (*Paragomphus*) wait at sandbanks, for example, as their nymphs live in the shifting sand that settles where faster and slower flowing waters intersect. With awl-shaped bodies and short legs, they almost swim among the grains. The antenna's last segment is thin and bent sideways, a feature present in the New World's unrelated "sand-swimmers" too (p. 104).

As in many pond dragons and stream damsels (compare p. 180), the male's perching behavior is linked to elaborate traits, although color seems less important. Most gomphids are quite similar, with brighter tones (blue eyes, russet club) often found in females too. Probably female preference is guided by the male claspers' shape and feel, as those vary vastly in the

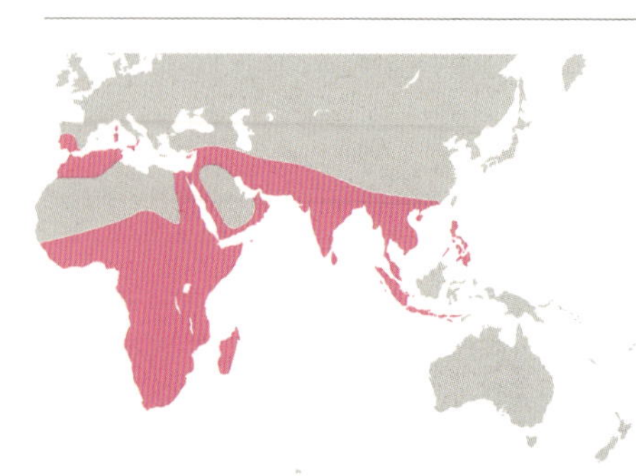

DIVERSITY
74 species of mostly running (but also standing and even temporary) waters from Africa and Madagascar to tropical Asia

TAXONOMY
Tribe Crenigomphini with genera *Cornigomphus* (2 species), *Crenigomphus* (6), *Libyogomphus* (5), *Nepogomphoides* (1), *Paragomphus* (48), *Tragogomphus* (3), and 9 tropical African and Malagasy species currently placed in *Onychogomphus* but that represent still unnamed genera (p. 101)

LEFT | A male of the Red-tipped Claspertail (*Onychogomphus aequistylus*) showing the deep rufous tail-end shared by many unrelated Malagasy odonates (p. 231).

BELOW | A male of Africa's Flapper Hooktail (*Paragomphus sabicus*) brandishing the abdominal foliations characteristic of many gomphids.

family (p. 106), facilitated by the diagnostically wide space between the eyes, where the male grasps the female.

Onychogomphinae possess the heftiest claspers. While having a similar distribution as Gomphinae, the species favor coarser substrates and thus often faster-flowing waters. Nymphs can be quite stocky and hide among gravel or detritus, or dig into fine material held between stones. Such spots offer good perches (like emerging rocks) and may be patchy (such as calm spots within a rapid stream), so can be controlled by the active males, holding their claspers aloft for all to see.

Tropical Asia's onychogomphines are most diverse (next profile). Almost 20 *Paragomphus* species occur there, but they must have originated in Africa, where 30 more varied species and the related African horntails (*Cornigomphus*, *Libyogomphus*, *Nepogomphoides*, *Tragogomphus*) and claspertails (placed in *Onychogomphus*) are found. This tribe is named for the talontails (*Crenigomphus*): their few species are so close that the older name *Crenigomphus* should probably be applied to all *Paragomphus* species.

PINCERTAILS AND KIN

As purported bringers of evil, European folklore links dragonflies to snakes. With its spotted pattern and headike club and beaklike claspers, a gomphid's raised abdomen is indeed quite serpentine. When Selys (p. 65) coined the name *Ophiogomphus* ("snake-bolt" in Greek, see p. 94) in 1854, he was inspired by *Aeschna serpentina*, a species known as the Green Snaketail (*O. cecilia*) today. Selys also created *Erpetogomphus* ("reptile-bolt"), and most of its species were subsequently given snaky names such as *E. boa*, *E. constrictor*, and *E. viperinus*.

The latter genus, known as American ringtails, is largely confined to Middle America, while all but a few Eurasian snaketail (*Ophiogomphus*) species occur in North America. These are the New World's only

ABOVE | Although gomphids are generally colored more to blend in than stand out (photo opposite), the bold onychogomphines often have bright green thoraces and in rare cases such as this male Blue-faced Ringtail (*Erpetogomphus eutainia*) from Texas even some blue.

LOWER RIGHT | Many gomphids raise their massive claspers high, but few also have such a weird humped profile as this male Camel-backed Grabtail (*Lamelligomphus camelus*) from Vietnam.

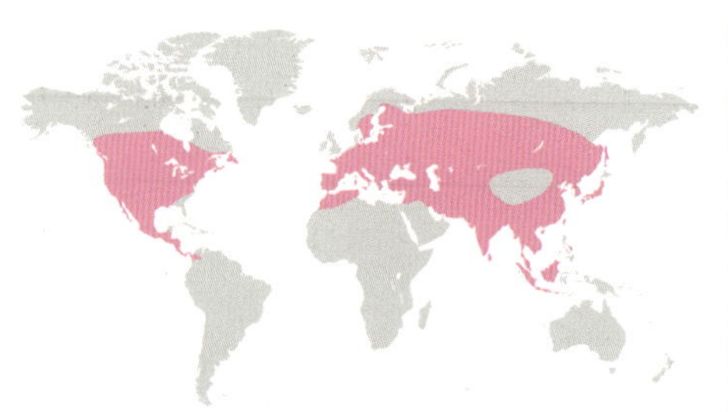

DIVERSITY
170 species of mostly running waters from tropical Asia across Europe to North and Central America

TAXONOMY
Tribe Onychogomphini with genera *Acrogomphus* (4 species), *Amphigomphus* (3), *Borneogomphus* (1), *Erpetogomphus* (24), *Lamelligomphus* (16), *Megalogomphus* (12), *Melligomphus* (11), *Nepogomphus* (3), *Nihonogomphus* (18), *Nychogomphus* (8), *Onychogomphus* (24), *Ophiogomphus* (26), *Orientogomphus* (7), *Perissogomphus* (1), *Phaenandrogomphus* (7), and *Scalmogomphus* (5)

Onychogomphinae, however, with most of their relatives (placed in the tribe Onychogomphini) restricted to the warmer parts of Eurasia (p. 99). All characterized by impressive male claspers, many were once called *Onychogomphus*, meaning "claw-bolt." The lower clasper's long branches typically run parallel, rather than diverging as in most Gomphinae.

As research progressed, most species were split off into other genera. Even today, not all dragonflies called *Onychogomphus* (see photo p. 23) are near that genus's type species, the Small Pincertail (*O. forcipatus*), and its close relatives in Europe and adjacent north Africa and west Asia (p. 98). As popularization progresses, ever more clasper-inspired common names are introduced too.

Nihonogomphus and *Scalmogomphus* still go by the name pincertail, but there are clawtails (*Acrogomphus*), pliertails (*Amphigomphus*), grabtails (*Lamelligomphus* and *Melligomphus*), sabretails (*Megalogomphus*), nippertails (*Nepogomphus*), and tongtails (*Nychogomphus*) now as well.

Even brackettails (*Orientogomphus*) and scrawnytails (*Phaenandrogomphus*) have intimidating claspers. *Borneogomphus* has yet to get a name, while the Himalayan Snaketail (*Perissogomphus stevensi*) is very close to *Ophiogomphus*.

ABOVE | Found on stony rivers in the dry landscapes of the Western Mediterranean, the Faded Pincertail (*Onychogomphus costae*) has camouflaging colors.

FLANGETAILS, TIGERS, AND KIN

Although flowing-water species disperse more poorly (p. 12), gomphids occur almost globally. They are absent from oceanic islands, however, and scarce on large islands like Madagascar and the Greater Antilles (compare pp. 180–1). Only single species occur on New Guinea, Sulawesi, and the islands in-between: all are flangetails (*Ictinogomphus*), placed in the tribe Lindeniini of the subfamily Lindeniinae.

These groups are named for the Bladetail (*Lindenia tetraphylla*), a 2¾ to 3⅛ in (7 to 8 cm) sand-colored dragonfly that breeds in large bodies of standing or slow-flowing water in rather arid regions (**blue on map**). Populations are patchy,

TOP LEFT | This female Bladetail (*Lindenia tetraphylla*) is named for Pierre Léonard Vander Linden, who first described it in 1825, and the four leaflike flaps on its club.

LOWER LEFT | The scientific name of the fearsome genus *Ictinogomphus*, including this Australian Flangetail (*I. australis*), means "hawk-bolt."

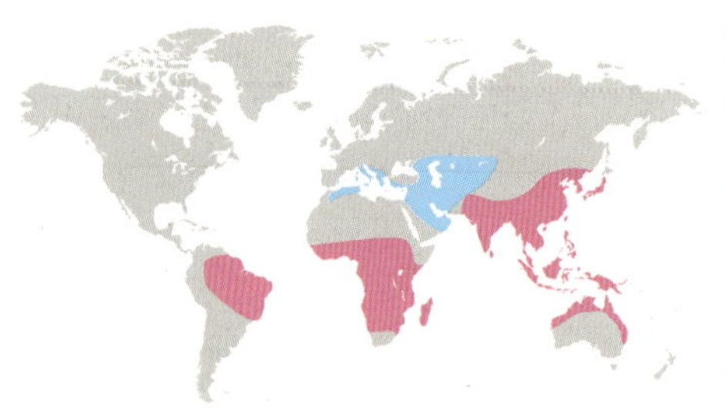

DIVERSITY
54 species in wide range of often larger and/or standing waters in warm parts of the world

TAXONOMY
Tribe Lindeniini (sometimes known as Ictinogomphini of subfamily Ictinogomphinae) with genera *Cacoides* (1 species), *Diastatomma* (6), *Gomphidia* (21), *Gomphidictinus* (3), *Ictinogomphus* (20, including *Sinictinogomphus*), *Lindenia* (1), and *Melanocacus* (2); African *Gomphidia* species should probably be separated as *Africogomphidia*

but its adults are the only gomphids that migrate, and they can establish rapidly and numbers become huge, for instance at new reservoirs. Mature adults can be almost black and even become pruinose, probably to protect them from solar radiation in their exposed environments (p. 65).

Ictinogomphus males perch on prominent stakes over marshes, ponds, lakes, and sluggish rivers from Africa to Japan and Australia. The big-leafed abdomen is held high, although much time is spent flying about aggressively. In South America, *Cacoides latro* looks and acts so similarly that it is considered a recent offshoot, as may also be the two poorly known *Melanocacus* species. Although all other species occur only in the Old World, this threesome makes Lindeniini the only gomphid tribe found throughout the tropics.

Aside from size, adults are recognized by their dense venation and diagnostic thick spines below the wing bases. Nymphs are equally distinctive, having short broad abdomens (almost circular from above) with a high ridge, somewhat like limpets. With their rather long legs and wide bodies, they sprawl among coarse organic material, rather than burrowing like most gomphids.

These nymphs probably survive anywhere with enough food and cover. Together with their size, this may have allowed the species to vault into a pole predatory position even in standing waters (compare p. 81) and across oceans. Of the tribe's stream and river species, tropical Asia's tigers (*Gomphidia* and *Gomphidictinus*) are near *Ictinogomphus*, suggesting that genus evolved there. The hoetails (*Diastatomma*) and three fingertails (currently in *Gomphidia* too) are more distinctive, however, suggesting the tribe as a whole arose in tropical Africa.

ABOVE | A male Western Hoetail (*Diastatomma gamblesi*) in Ghana.

GOMPHIDAE—LINDENIINAE—*APHYLLA, PROGOMPHUS, ZONOPHORA*, AND RELATED GENERA

SANDDRAGONS, FORCEPTAILS, AND KIN

Almost four-fifths of Lindeniinae are restricted to the New World. Superficially, adults there are quite like their Old World counterparts, slender with comparatively dense venation. Males typically possess long fingerlike upper claspers (often bent toward each other like forceps), while the lower clasper is rather simple and often reduced.

The species are much more varied in size and habits, however, their nymphs showing the lifestyles found in other subfamilies elsewhere. Sanddragons (*Progomphus*, Progomphini) burrow rapidly into sand, for example, having similar torpedo-shaped bodies (and sometimes even rod-tipped antennae) to *Paragomphus* (p. 98). Males may guard sandbars in rivers or beachlike bays in streams.

Forceptails (*Aphylla* and *Phyllocycla*) are tropical America's deep-burrowers (compare pp. 94–5), the nymphs' stretched final abdomen segment serving as a breathing-tube. *Aphylla* may frequent sloughs and ponds, but the generally smaller *Phyllocycla* only streams and rivers. The closely related American leaftails (*Phyllogomphoides*) at most burrow shallowly in mud and detritus in calm spots of flowing waters. The hardly known *Gomphoides*, *Idiogomphoides*, and *Peruviogomphus* complete the tribe Gomphoidini.

Very little is also known about the 19 remaining species, grouped in Zonophorini. Calipertail nymphs (*Zonophora*) may hide among detritus and roots, while American pincertails (*Desmogomphus*) and pegtails (*Perigomphus*) probably burrow shallowly in sand and gravel of small forest streams. Those of *Diaphlebia* are unknown, while *Anomalophlebia*, described from a single female, may be synonymous with one of these genera.

LEFT | A male Four-striped Leaftail (*Phyllogomphoides stigmatus*) from Texas, USA, showing the foliations for which it is named.

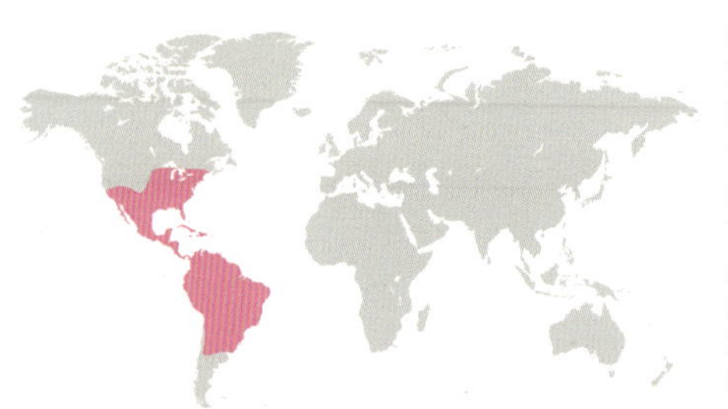

DIVERSITY
201 species in all kinds of running (and some standing) waters in the warmer parts of the Americas

TAXONOMY
Tribe Gomphoidini with genera *Aphylla* (24 species), *Gomphoides* (4), *Idiogomphoides* (3), *Peruviogomphus* (3), *Phyllocycla* (31), and *Phyllogomphoides* (47); Progomphini with *Progomphus* (70); and Zonophorini with *Anomalophlebia* (1), *Desmogomphus* (3), *Diaphlebia* (2), *Perigomphus* (3), and *Zonophora* (10)

VICETAILS AND KIN

Representing about a quarter of gomphid species, Lindeniinae appear to have separated before the rest of the family began to diversify. Within this subfamily, the 13 species forming the tribe Hemigomphini are most distinct and can thus be regarded as the evolutionarily most isolated gomphids.

Seven vicetails (*Hemigomphus*) occur in the temperate forests of eastern Australia and the north's tropical rainforests, inhabiting streams and rivers with sand, gravel, and cobbles. The Armourtail (*Armagomphus armiger*; photo p. 21) and Pinchtail (*Odontogomphus donnellyi*) are restricted to the far southwest and northeast of this island continent, respectively.

The Atlantic Vicetail (*Praeviogomphus proprius*) only occurs near Rio de Janeiro in Brazil, while three Patagonian vicetails (*Neogomphus*) live in the temperate Valdivian Forests in central Chile. The hairy males perch on rocks and bushes above cold lakes, rivers, and streams with gravel and pebbles.

Found mostly in south-temperate habitats, hemigomphines present a classic Gondwana distribution (p. 111). Combined with the subfamily's principally South American diversity (see opposite) and the possibly African origin of Lindeniini (p. 103), this suggests that Lindeniinae evolved on that southern supercontinent, while Gomphinae (p. 94) and their nearest relatives (p. 96), as well as Onychogomphinae (pp. 98–101), were centered in the northern supercontinent, Laurasia (p. 111).

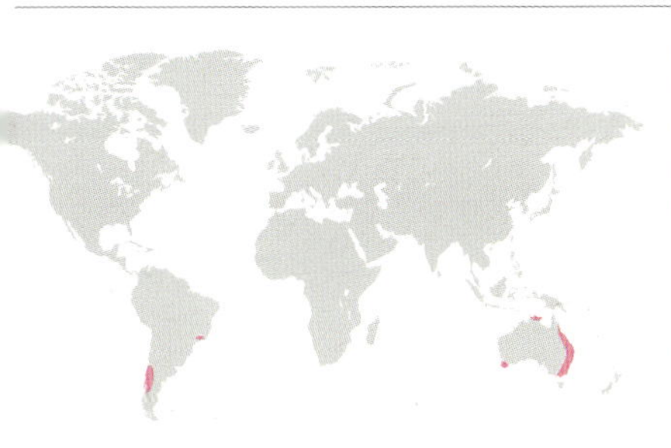

DIVERSITY
13 species of streams, rivers, and lakes in Australia and southern South America

TAXONOMY
Tribe Hemigomphini with genera *Armagomphus* (1 species), *Hemigomphus* (7), *Neogomphus* (3), *Odontogomphus* (1), and *Praeviogomphus* (1); *Eogomphus neglectus* was placed here too, but nothing is known beyond its description in 1930 from China

ABOVE TOP | Restricted to southwest Australia, the Armourtail (*Armagomphus armiger*) is the most unusual looking species in this group.

ABOVE BOTTOM | The hairy male Patagonian Vicetail (*Neogomphus edenticulatus*), having just emerged from a lake in the Argentinian Andes, is well-adapted to its cool environment.

KNOBTAILS, SNOUTTAILS, AND SHELFTAILS

DIVERSITY
55 species of forest streams from Mexico to Brazil

TAXONOMY
Tribe Archaeogomphini with genus *Archaeogomphus* (7 species); Cyanogomphini with *Agriogomphus* (4), *Brasiliogomphus* (1), *Cyanogomphus* (3), *Ebegomphus* (5), and *Tibiagomphus* (2); and Epigomphini with *Epigomphus* (33), all placed tentatively in subfamily Epigomphinae

ABOVE | A male of the Fork-tipped Knobtail (*Epigomphus quadracies*) in Panama showing its massive abdomen tip and claspers.

RIGHT | A male of the Rio Snouttail (*Archaeogomphus infans*) in Brazil, an unusally fragile gomphid.

The previous profiles cover 83 percent of all gomphids known, placed in just a few rather well-defined subfamilies. These are hard to diagnose, though, even the important male claspers seeming too complex and varied to characterize. Lindeniines typically have elongated forceps, thus extending along the body's main axis (p. 102), while onychogomphine claspers can be quite high, so often developed vertically (p. 101). Gomphines frequently have splayed appendages, directed more perpendicularly to the body (p. 92).

The remaining 171 species fall into seven groups, mostly classified as tribes of the subfamily Epigomphinae. Each is morphologically, genetically, and geographically so isolated that they may ultimately prove to be separate subfamilies. Interestingly, the three limited to the American tropics each have a unique configuration of the male's abdomen tip too.

In the knobtails (*Epigomphus*) of the tribe Epigomphini the "tail" is not clubbed, but enlarged right at the end, presumably holding substantial musculature for the males' large, complex claspers. Unlike other gomphids, some females settle to insert their eggs into wet mud. Nymphs live in the mix of sand, mud, and detritus in small forest streams.

The snouttails (*Archaeogomphus*) of Archaeogomphini may be the oddest gomphids. At most 1⅓ in (3.5 cm) long, the male's terminal abdomen segment is drawn out into a snout-like point and bears two recurved hooks at its base. These might take on the role of the highly reduced claspers. The smooth and spiny nymphs can be almost translucent green, looking more like sprawling libellulids than burrowing gomphids, having been found clinging to vegetation hanging into a shallow stream.

The shelftails of Cyanogomphini are placed in five very similar genera. The upper claspers are implanted on a shelflike overhang of the abdomen tip, with tusklike hooks curving down and back into the body. The broad and spiny nymphs are equally weird with an enlarged central spine on their second-last abdominal segment. They hide (or burrow shallowly) among organic material.

GOMPHIDAE—REMAINING OLD WORLD GROUPS

THORNTAILS, FAIRYTAILS, SCISSORTAILS, HUNTERS, AND KIN

BELOW | The common name of Madagascar's glyphtails (*Isomma*) refers to the scientific name of the Red-tipped Glyphtail (*I. hieroglyphicum*) and its intricate male claspers. Note the deep reddish tip characteristic of many Malagasy odonates (p. 231).

The Old World too harbors several rather isolated gomphid groups that may represent distinct subfamilies (see previous profile). Aside from the few lindeniines arriving recently from Asia (p. 103) or with distant relatives in Patagonia (p. 105), the 26 austrogomphines are Australia's only gomphids (**dark pink on map**). They are quite unremarkable, though: neither nymph nor adult hunters (*Austroepigomphus*, *Austrogomphus*, *Zephyrogomphus*) might stand out immediately in Africa or Eurasia. In the dragons (*Antipodogomphus*), however, both life-stages have the last two abdomen segments extended, thus representing Australia's deep-burrowers (compare pp. 94–5).

The last abdomen segment in male adults of southern Africa's thorntails (*Ceratogomphus*; photo p. 4), in Phyllogomphinae, bears a distinctive ridge that is drawn forward into a bladelike thorn. The segment is enlarged in the African leaftails (*Phyllogomphus*; photo p. 93), the arched ridge adorned with fine denticles, like

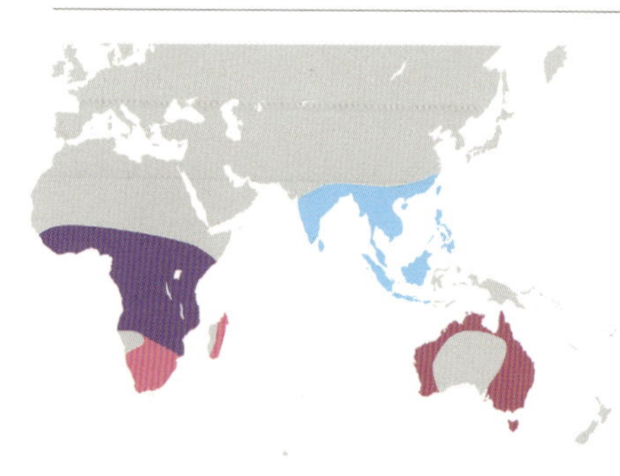

DIVERSITY

116 species of running waters (occasionally larger standing waters) in tropical Asia, Africa, Madagascar, and Australia, including Tasmania

TAXONOMY

Genera *Ceratogomphus* (2 species), *Isomma* (3), and *Phyllogomphus* (11) placed in subfamily Phyllogomphinae; tribes Lestinogomphini with *Lestinogomphus* (10), Microgomphini (here including Leptogomphini) with *Heliogomphus* (21), *Leptogomphus* (25), *Microgomphus* (17), and possibly *Davidioides* (1); and Austrogomphini with *Antipodogomphus* (6), *Austroepigomphus* (3), *Austrogomphus* (15), and *Zephyrogomphus* (2) are assigned very tentatively to Epigomphinae

a crested helmet. The imposing males (about 3⅛ in/8 cm long) perch conspicuously at sunny streams, rivers, and lakeshores across equatorial Africa, their abdomens raised to show off this helmet and the huge flaps on their third-last segment. While the rather standard-looking *Ceratogomphus* nymphs burrow shallowly in calm waters, the huge *Phyllogomphus* nymphs have strong legs to dig deep into muddy bottoms and a long abdomen, its final segment forming a short breathing-tube.

Hiding among organic material in the same waters, fairytail (*Lestinogomphus*) nymphs have similar snorkels as *Phyllogomphus*, but are tiny with longish legs. Measuring just over 1½ in (4 cm), with narrow wings and slender abdomens, the adults recall *Lestes* damselflies, hence the genus name (p. 240). As in the nymphs, the last segment is elongated, bent down sharply at rest.

ABOVE | Male of Australia's Pale Hunter (*Austrogomphus amphiclitus*).

RIGHT | A male of the Congo Fairytail (*Lestinogomphus congoensis*) in Gabon characteristically bending its final segment down.

Although lestinogomphines probably occur on most running waters in tropical Africa, half the species were named only recently.

Adult scissortails (*Microgomphus*) are similarly small and shy. With wide, very flattened bodies looking like the leaf-litter they hide in, the tiny, dark nymphs are easily found in forested streams. Besides five species from tropical Africa, twelve are currently known from Asia, all dark with characteristic shear-like claspers.

The Asian grappletails (*Heliogomphus*) have been put with *Microgomphus* in Microgomphini, but also with the slendertails (*Leptogomphus*) in Leptogomphini, so they might be united as one group of at least 63 species (**blue and purple on map**). The two other genera (both confined to Asian forest streams) have equally slender, dark, and secretive adults, while the nymphs are similarly leaflike.

The Syrandiri Clubtail (*Davidioides martini*) occurs only in India's Western Ghats. Having been associated with *Davidius* (p. 97) and Onychogomphini (p. 100), but the present group as well, its affinities are still mysterious. Discovery of its nymph is eagerly anticipated, therefore!

FAMILY: PETALURIDAE

PETALTAILS

Petalurid nymphs are the shrews of the dragonfly world, emerging from burrows at night to hunt terrestrial invertebrates. Dug into waterlogged soil, their lairs always remain wet. With an enlarged end-chamber, multiple side-chambers, and tunnels up to 30 in (75 cm) deep, the dens of the Australian petaltails (*Petalura*) and New Zealand's giants (*Uropetala*) are most impressive. The Chilean Petaltail (*Phenes raptor*) in the Patagonian Andes, by contrast, lives among leaves and woody debris in shallow seeps or, in higher rainfall areas, even on the damp forest floor.

While the dragons treated thus far mostly lay eggs freely in water (p. 93), the remaining groups and damsels retain the ovipositor with which Odonata first evolved (p. 18), boring eggs into plant tissue and other substrates such as mud. Petaltail females are unique in combining this apparatus with widely spaced eyes, while males have absurdly large (indeed petal-like) claspers. We will first be struck by their size, however. Among the biggest odonates, the Giant Petaltail (*P. ingentissima*) nymph is almost 2¾ in (7 cm) long; the adult female can measure 5 in (12.5 cm) with a nearly 6¾ in (17 cm) wingspan. Living mostly in temperate environments, petalurid adults often bask on rocks or tree trunks to warm their massive dark bodies, even settling purposefully on light-clothed humans!

ABOVE LEFT | A male of New Zealand's Mountain Giant (*Uropetala chiltoni*) showing the petal-like claspers on its "tail."

OPPOSITE | North America's spectacular Gray Petaltail (*Tachopteryx thoreyi*) typically lands flat on light-colored surfaces such as tree trunks to warm up.

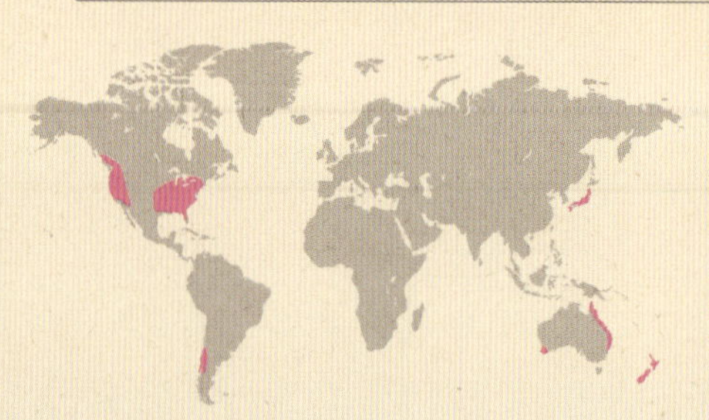

DIVERSITY
10 species of bogs, seeps, streambanks, and other permanently wet ground in North America, Patagonia, Japan, Australia, and New Zealand

ADULT HABIT
Mostly very large and dull-colored dragonflies with widely spaced eyes, extremely long stigmas, and ovipositor at female's abdomen tip

TAXONOMY
Subfamily Petalurinae with genera *Petalura* (4 species), *Phenes* (1), and *Uropetala* (2); Tachopteryginae with *Tachopteryx* (1) and *Tanypteryx* (2)

Some have speculated that the first dragonflies were burrowers too. Petalurid-like wings with dense venation and extended stigmas have indeed been found as fossils from the Jurassic and Cretaceous around the world. Three northern species, moreover, are genetically well-separated from the seven in the distant south, suggesting they are the only survivors of a once globally dominant group.

Confined to Honshu and Kyushu, the Japanese Petaltail (*Tanypteryx pryeri*) has simple burrows up to 9½ in (24 cm) deep. Across the Pacific, the Black Petaltail (*T. hageni*) rarely goes beyond 4 in (10 cm), but is also the smallest giant, adults not exceeding 2⅓ in (6 cm). Found on North America's other coast, the Gray Petaltail (*Tachopteryx thoreyi*) is larger but frequents pits that are at most half an inch (1.5 cm) deep and 1½ in (4 cm) wide in wet mud, usually under leaves on ground kept wet by seepage.

The family's north–south divide aligns nicely with the breakup of Pangea, beginning roughly 180 to 140 mya, with the resulting supercontinents of Gondwana and Laurasia eventually fragmenting further to form today's continents. New Zealand has been isolated for 80 million years, for instance, and it is hard to imagine *Uropetala* flying there. So it does appear that petalurids have remained where they are for a long time. Paleontologists have concluded that ancestral dragonflies lived in open water, however, and no fossils actually match extant petalurids exactly. While the burrowing giants may have ranged more widely (on Antarctica, for example, p. 128), therefore, it seems likeliest that their extreme specialization kept them localized, but also allowed petalurids to survive for ages under the right conditions.

Soggy bog-like habitats persist best in cool wet places, like the temperate and coastal regions inhabited today, which have changed little since the Jurassic. A burrow, meanwhile, limits predation risk from usual nemeses such as fish and other odonates, but also competition with other dragonflies, particularly as threats on the surface like fire and drought can (to a degree) be escaped underground. The deep-burrowers probably also dig new tunnels and chambers to adjust to change in the water table.

As these habitats are always patchy and growing so large may take up to ten years, the species must be vulnerable to the rapid rates of development and climate change today, however. Currently, only Australia's Coastal Petaltail (*P. litorea*) may be at risk, ranked as Near Threatened on the IUCN Red List, although data on the Western Petaltail (*P. hesperia*) are insufficient to be certain.

FAMILY: AESHNIDAE
DARNERS AND HAWKERS

BELOW | Unlike most dragonflies species, which perch frequently, aeshnids such as this male Blue-faced Darner (*Coryphaeschna adnexa*), fly about for long periods when active.

As the biggest dragons from Arctic ponds to Amazonian streams, aeshnids have little to be ashamed about. Many have nonetheless argued that *Aeshna* comes from the Greek for "shame," *aischyne*, subsequently appending "*aeschna*" to most genus names in the family. Johan Fabricius, who picked *Aeshna* 250 years ago, always stuck to his spelling, though. In popular parlance, these dragonflies are generally known as darners in the Americas and Australia, but hawkers elsewhere. After Libellulidae (p. 34), Aeshnidae is the main dragonfly family of standing waters. With Austropetaliidae (p. 128), they separated before the other extant dragonfly lineages diverged, taking quite a different approach to colonizing these waters (compare p. 238). Aside from mostly being larger, for example, adults have rather long abdomens and wings but short legs. They rest hanging and often hidden, therefore, yet remain aloft for extended periods once active. With this ability, males patrol a chosen beat

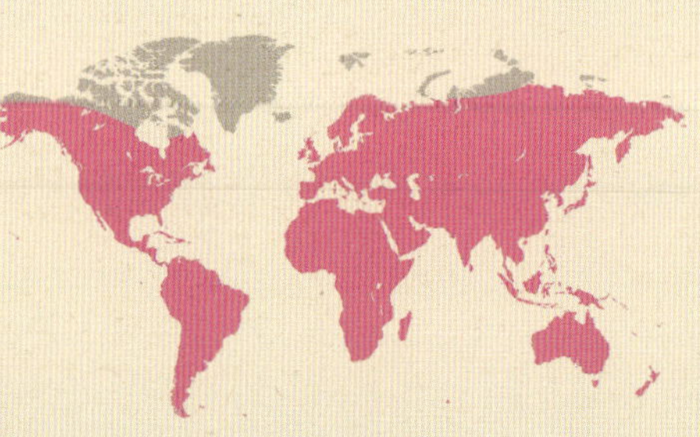

DIVERSITY
495 species at almost any water where odonates occur, from cold streams to drying pools

ADULT HABIT
Largest dragonflies in many parts of the world, especially at standing water, with comparatively long bodies, wings, and stigmas, large eyes, and habit of flying continuously when active

TAXONOMY
At least 4 rather well-defined subfamilies are recognized, of which the most diverse and widespread is treated first, divided into 3 groups of ecologically similar genera; 1 group of genera with similar ecology but uncertain affinities is treated together (p. 124)

ABOVE | Female aeshnids such as this Paddle-tailed Darner (*Aeshna palmata*) in the USA must land to lay eggs.

or doggedly inspect every corner of a waterbody. Aided by perhaps the largest of all insect eyes, with up to 30,000 ommatidia (see photo p. 12), both sexes feed in sustained flight, often congregating to feast on small insects swarming in sheltered places or above the trees. They may take other dragonflies too, even those as large as themselves.

In stark contrast, females must land to lay eggs, as they are the only dragons with ovipositors in most places (compare pp. 86 and 93). Furtively moving along the waterside, they seek out the best possible spots to place them. The nymphs are comparatively long-bodied and short-legged too, clinging to vegetation, roots, or the underside of rocks or debris. With big eyes and a long mask, they ambush or stalk their often large prey—which may be small fish or tadpoles—and strike from a distance, as chameleons do, allowing them to attain their

LEFT | A male Fat-tailed Darner (*Rhionaeschna diffinis*), a species found mostly in the temperate Andes.

BELOW | Nymph of the Common Green Darner (*Anax junius*) from Texas, USA.

substantial size (up to 2¾ in/7 cm) relatively quickly. The nymphs are themselves vulnerable to predation too, even by their own species. Many have broad piebald crossbands when small, perhaps concealing them from visual predators. Aeshnids thus appear to favor situations in which they can quickly become the top dog. Maybe cannibalism even drove this evolution, as the risk of being eaten by larger siblings is lowest if all complete their lifecycle in one season.

The nymph's terminal appendages are often modified into a gutterlike siphon. Otherwise present only in the semiterrestrial petalurids (p. 110) and austropetaliids (p. 128), this may allow them to inhale air where oxygen (or even water) is scarce—for example, at the surface of warm, fetid pools or on the wet forest floor: habitats that perhaps have few competitors and predators, but lots of food.

While their distinct ecological strategy allowed aeshnids to occur wherever odonates do, from deserts to rainforests, as well as on most oceanic islands, their diversity and abundance are modest in many places. Under special conditions, however, their adaptations allow incredible numbers to develop. In shallow lakes in China, nymphs of the Eastern Lesser Emperor (*Anax julius*; p. 118) can be harvested commercially for human consumption, for example!

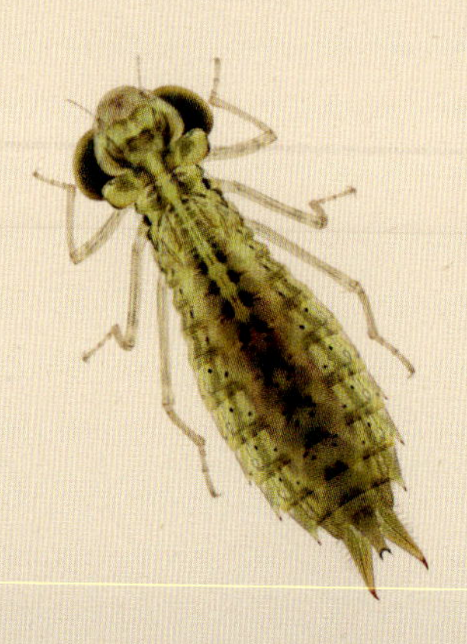

MOSAIC DARNERS AND HAWKERS

Among the marginal environments that aeshnids may dominate (see opposite) are the coolest the warm-loving odonates inhabit. *Aeshna* species can be abundant in bogs, fens, lakes, and slow-flowing waters in the Northern Hemisphere's temperate forests and adjacent tundra (see photo p. 113; **pink on map**). Adults often bask on trees, rocks, or bare ground, the mosaic of bright spots on their black abdomens allowing them to trap heat.

Although the American riffle darners (*Oplonaeschna*) and Springtime Darner (*Basiaeschna janata*) from rocky and wooded streams in northern Middle America and eastern North America have long been recognized, many distinct but related genera were separated only recently. Despite its plain brown body and much earlier flight season, Europe's Green-eyed Hawker (*Isoaeschna isoceles*) remained in *Aeshna* for two centuries!

This is especially true in the south (those genera **blue on map**), where cooler habitats are much more isolated (p. 78). *Rhionaeschna* occupies calm waters in South America's highlands and south, particularly the Andes and Patagonia (photo opposite); some species follow the mountains into North America as well, overlapping with *Aeshna*.

With stunningly all-red or all-blue males, *Andaeschna* inhabits streams in the equatorial Andes, while *Afroaeschna*, *Pinheyschna*, and *Zosteraeschna* occur mainly in upland and forest streams in Africa. *Adversaeschna* and the related *Oreaeschna* are found in various waters in upland and temperate areas of New Guinea, New Caledonia, Australia, and New Zealand.

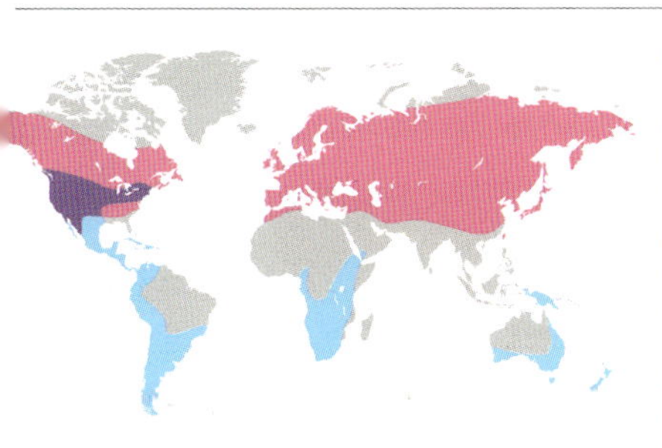

DIVERSITY
93 species found around the world in most kinds of freshwater habitats, but predominantly in cooler areas (mountains, temperate zone)

TAXONOMY
Genera *Aeshna* (29 species), *Adversaeschna* (1), *Afroaeschna* (1), *Andaeschna* (5), *Basiaeschna* (1), *Isoaeschna* (1), *Oplonaeschna* (2), *Oreaeschna* (2), *Pinheyschna* (6), *Rhionaeschna* (42), and *Zosteraeschna* (3)

ABOVE | A female Orange-headed Hawker (*Oreaeschna dictatrix*) ovipositing in a highland pool on the island of New Guinea.

DUSKHAWKERS AND ASSOCIATES

While many dragonflies emerge from the water at night, and some even migrate then, this time generally seems too dark and cold for odonates to hunt by sight and flight, despite the abundance of small insect prey. In an insect order associated with sun, the family Aeshnidae has pushed the limits of the day most successfully, however. Many temperate species (p. 115) feed particularly on balmy evenings, however, with twilight foraging preferred by most in the tropics. Large-eyed even by aeshnid standards, these crepuscular species may hunt at dawn or with rain, but mostly during the equatorial sunset's half-hour dusk, whizzing about wildly in clearings or hovering cautiously, low over the ground. They might even be active later as they come to light more frequently than other dragons. Often largely green or brown, adults hang inside dense vegetation (sometimes also under overhangs or in buildings) during the day, although males may search the tangles for mates.

Some hundred duskhawkers (called two-spined darners in the Americas) are placed in *Gynacantha*. While found in almost any tropical environment from Florida's cypress swamps to Australia's arid interior, where the Cave Duskhawker (*G. nourlangie*) seeks out shaded rock pools, most inhabit forest.

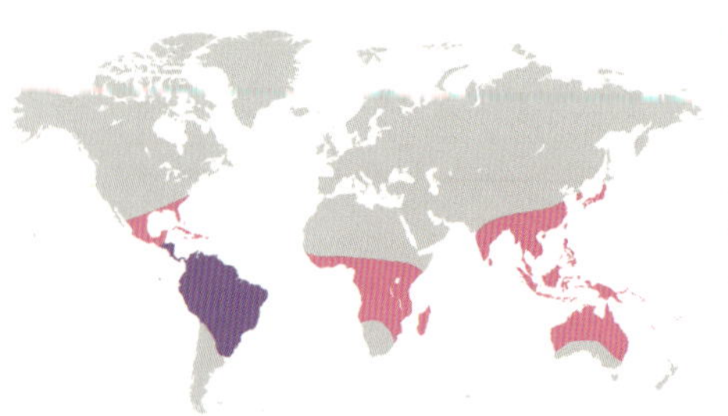

DIVERSITY

150 species of shady tropical habitats, mostly with smaller standing (and often temporary) waters

TAXONOMY

Genera *Agyrtacantha* (6 species), *Austrogynacantha* (1), *Gynacantha* (100), *Heliaeschna* (11), *Plattycantha* (3), and *Triacanthagyna* (9) seem close (**pink and purple on map**), as *Neuraeschna* (15) and *Staurophlebia* (5) may also be (**purple only**)

Untangling this heterogeneous genus depends on clarifying the relationships of various similar genera, however. While placed in *Heliaeschna*, for example, Asia's nighthawkers are unlike that otherwise African genus. The Australasian genera *Agyrtacantha*, *Austrogynacantha*, and *Plattycantha* and the tropical American three-spined darners (*Triacanthagyna*) seem close to *Gynacantha* too.

The species not only seek out the fringes of the day to feed, but also push the limits to breed, favoring some of shadiest, smallest, and shortest-lived habitats. Nymphs attain the greatest densities in temporary waters (rain pools, muddy stream margins, even tree holes) sheltered by forest or thickets. After emerging, the cryptic adults may have to wait for months to breed, lurking in the undergrowth near dry flood-forest or streambeds.

Some species oviposit in dried-out pans before they refill. *Gynacantha* translates from Greek as "woman-thorn," referring to two claw-like spines on the female's abdomen tip. These may be used to gauge, grip, or perhaps scrape the leaf-litter, fallen wood, or hardened mud where the eggs are placed. As names such as *Polycanthagyna* (p. 119) , *Tetracanthagyna* (p. 120) and those above indicate, the spines' number, character, and function vary widely.

Although probably related to the others, the vesperdarners (*Neuraeschna*) and megadarners (*Staurophlebia*) of tropical America's lowland forests favor running waters more. *Neuraeschna* species are dark and crepuscular like duskhawkers, but the formidable males (up to 4 in/10 cm) of the related *Staurophlebia* are bright green and orange, and patrol long distances in full sun over clearings, streams, and rivers.

LEFT | With often large appendages that may assist in oviposition, *Gynacantha* females, such as this Bar-sided Darner (*G. mexicana*) from the American tropics, are easily mistaken for males.

RIGHT | While considered related to crepuscular groups, the brightly colored megadarners (*Staurophlebia*) of tropical America, such as this male Magnificent Megadarner (*S. reticulata*), are more diurnal, recalling *Anax* (p. 118).

EMPERORS AND ASSOCIATES

While most aeshnid adults are quite dark or even dull, those of open and pondlike habitats rival the libellulids found alongside them in gaudiness: almost every large dragonfly that flies continuously and is extensively bright red, blue, and/or green is an aeshnid, most often an emperor (*Anax*).

All species (pink and purple on map) are large, powerful fliers. Among the biggest odonates, Africa's 4¾ in (12 cm) Black Emperor (*A. tristis*) appears after rain. Masses of Vagrant Emperors (*A. ephippiger*) emerge from temporary pools in the warm parts of Africa and Eurasia. Searching for new habitat, it colonized the Caribbean and reached Iceland (where no odonates breed) several times.

Each year, Common Green Darners (*A. junius*) born in Mexico, the Caribbean, and southern USA fly as far as Canada to breed, their offspring returning south in the fall. Strays have reached Europe and eastern Siberia, but must also have given rise to the Hawaiian Darner (*A. strenuus*).

Although most *Anax* breed in sunny, still waters rich in vegetation (photo p. 18); some, like the Magnificent Emperor (*A. immaculifrons*) from Asia, inhabit streams and rivers with sometimes abundant shade, their great size perhaps providing the competitive advantage to colonize water that's warm and productive enough (compare p. 103).

The genus is scarcer in the American tropics (**purple on map**), other genera such as *Staurophlebia* (p. 117) being the large, colorful, sun-loving aeshnids there. The pilot and malachite darners (*Coryphaeschna* and *Remartinia*) inhabit ponds, swamps, and slow-flowing waters in forested lowlands, while the closely related emerald darners (*Castoraeschna*) favor rivers and streams.

All aeshnids discussed so far are closely related. *Anax* and the complex with *Aeshna* (p. 115) seem especially close, as seen in the name and morphology

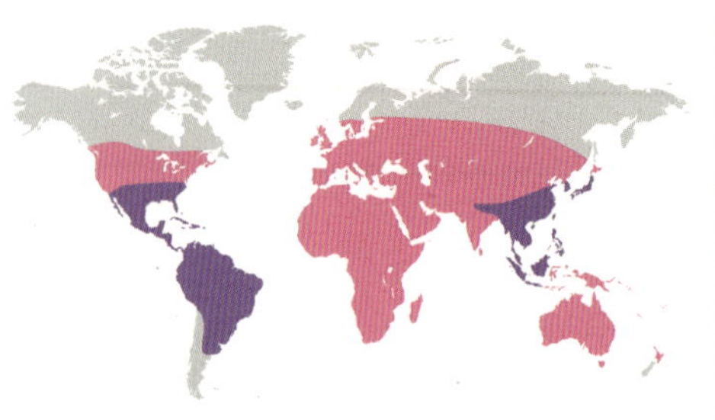

DIVERSITY
70 species inhabiting a broad range of standing and flowing waters in the world's warmer regions

TAXONOMY
Genera *Amphiaeschna* (1 species), *Indaeschna* (2), and *Polycanthagyna* (3) seem close, as do *Castoraeschna* (9), *Coryphaeschna* (8), and *Remartinia* (4); *Anax* (36) and *Anaciaeschna* (7) seem nearer *Aeshna* and relatives (p. 115)

ABOVE | Unlike other *Anax*, this male Magnificent Emperor (*A. immaculifrons*) from southwest Asia looks and behaves remarkably like other dragonflies that breed in (often shaded) rivers and streams, such as the goldenrings (p. 86).

of the genus *Anaciaeschna*. Ecologically, these evening-hawkers are nearer duskhawkers (p. 116), mostly flying in low light and breeding in (often forested and coastal) standing waters in tropical Africa, Asia, and the Pacific.

Six distinctive Asian aeshnines (**also purple on map**) are adapted to shady forest pools, typically filled by rain or rising streams. Males hang in the branches above them, the tigerhawkers (*Polycanthagyna*) being big and black with bright blue or green eyes and bold markings in blue, green, yellow, and red. Almost 4 in (10 cm long, the junglehawkers (*Indaeschna*) from Sundaland and the Philippines may even breed between tree buttresses (compare p. 156). Found near streams, Java's *Amphiaeschna ampla* may have similar habits.

ABOVE | A pair of Common Green Darners (*Anax junius*) ovipositing in the USA; the widest-ranging *Anax* species are among the few aeshnids (all wandering species) to do so with the male and female in tandem.

RIGHT | The spectacular male of the Junglehawker (*Indaeschna grubaueri*) in peninsular Malaysia.

AESHNIDAE—BRACHYTRONINAE

RIVERHAWKERS, SWAMP DARNERS, AND KIN

The female Great Riverhawker (*Tetracanthagyna plagiata*) has the widest wingspan of any dragonfly (6⅔ in/17 cm) and may be the heaviest odonate, even bulkier than *Petalura* (p. 110). All adult riverhawkers are most active in twilight. They are largely dull brown, often with dark rays in the wings and, in females, broad subapical crossbands as well. Confined to (rapid) streams and rivers in Southeast Asia (**blue on map**), mostly in forest, the nymphs climb above the surface at night to snatch small fish that pass beneath them.

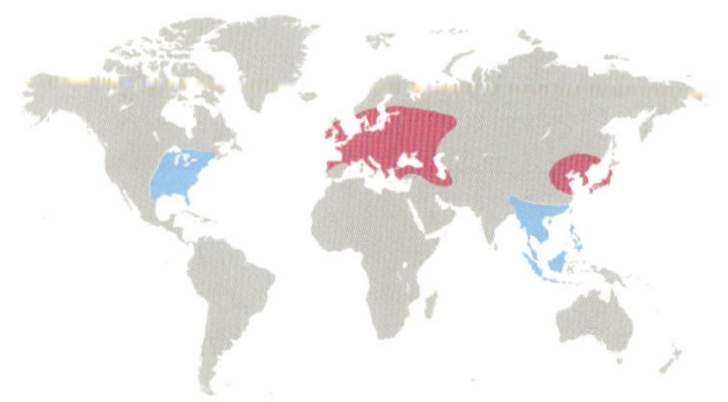

DIVERSITY

5 species in calm well-vegetated waters in temperate Eurasia and eastern North America, and 5 in (forested) streams and rivers in Southeast Asia

TAXONOMY

Genera *Brachytron* (3 species), *Epiaeschna* (1), *Nasiaeschna* (1), and *Tetracanthagyna* (5)

ABOVE | Europe's small Hairy Hawker (*Brachytron pratense*) is the first aeshnid there to appear in spring.

TOP RIGHT | The female Swamp Darner (*Epiaeschna heros*) is one of North America's largest aeshnids.

LOWER RIGHT | Great Riverhawker (*Tetracanthagyna plagiata*) laying eggs inside a log in Malaysia.

All brachytronine males have a cylindrical abdomen base, which is never constricted and waist-like, and a keel running over the back of the nymph's abdomen tip. Nonetheless, the other genera appear very different, breeding in standing (at most slow-moving) water in temperate regions. Western Eurasia's Hairy Hawker (*Brachytron pratense*), for example, recalls the *Aeshna* species found there. It flies in spring, however, well before most other aeshnids appear. Being smaller, stouter, and downier, the male patrols low along marshy margins, constantly entering the vegetation for a closer look.

Two similar species from northeastern Asia, long known as *Aeschnophlebia* (but see p. 124), were recently placed in *Brachytron* too (**pink on map**), as was also suggested for two species of various wooded habitats from North America (**also blue on map**). Named for its projecting face, the Cyrano Darner (*Nasiaeschna pentacantha*) flies over open water for a longer season, however. The Swamp Darner (*Epiaeschna heros*) seems to forage primarily above the trees, perhaps due to its size, being more than 3 in (8–9 cm) long rather than below 2 in (5–7 cm) as the foregoing species.

AUSTRALIAN DARNERS

Like many of its dragonflies, aeshnids in Australia evolved largely in isolation, not even crossing to New Guinea (compare p. 108). Including perhaps the family's strangest and most varied species, Australian experts have even ranked the subfamily Telephlebiinae as a distinct family. All inhabit running waters, although full-grown Terrestrial Evening Darner (*Antipodophlebia asthenes*) nymphs have been found under damp logs meters away from rainforest streamlets. Perhaps they mostly live like *Telephlebia* evening darners, among wet leaf-litter at stream margins and in waterfall spray. With oversized heads and scrawny bodies and legs, these nymphs seem even weirder than the adults. Those are marked boldly but only in earth tones, with dark wing patches and bright stigmas that may make them cryptic to predators among tangled vegetation in daytime, yet conspicuous to each other while flying about at dusk.

The very spiny and strong-legged nymphs of the cascade and Australian riffle darners (*Spinaeschna* and *Notoaeschna*) are the extreme opposite, clasping to the underside of rocks in fast-flowing water, like the similar-looking *Zygonyx* in Africa and Asia (p. 54). Females descend underwater, even in rapids, to lay their eggs. While common in damselflies, which easily close

DIVERSITY
36 species at a wide range of flowing waters in eastern Australia and Tasmania, plus 1 in the southwest

TAXONOMY
Genera *Acanthaeschna* (1 species), *Antipodophlebia* (1), *Austroaeschna* (20), *Austrophlebia* (2), *Dendroaeschna* (1), *Dromaeschna* (2), *Notoaeschna* (2), *Spinaeschna* (2), and *Telephlebia* (6)

TOP LEFT | Ocher-tipped Darner (*Dromaeschna weiskei*) male cruising over a rainforest stream.

LOWER LEFT | Male of the Southern Evening Darner (*Telephlebia brevicauda*).

OPPOSITE | Male of the Southern Giant Darner (*Austrophlebia costalis*).

their wings, this is almost unique among dragonflies. Unlike other aeshnids, moreover, they have weak ovipositors, and attach their eggs to rocks rather than placing them inside soft substrates. With its pug-like circular face, the Wide-faced Darner (*Dendroaeschna conspersa*) takes an intermediate position, sometimes submerging completely but always laying in dead wood.

Generally crepuscular, the group's adults can be very elusive. Like the mammal providing its common name, the Thylacine Darner (*Acanthaeschna victoria*) was even long deemed extinct. It proved highly specialized, however, living in blackwater streams that are oxygen-poor, acidic, periodically dry, and largely confined to the coastal plain and therefore mostly lost to human sprawl.

Owing to a marked radiation centered on the Australian Alps over half the group's species are Australian darners (*Austroaeschna*). The two rainforest darners (*Dromaeschna*) were long included in that genus, but are limited to tropical streams and among the more colorful telephlebiines. Giant darners (*Austrophlebia*) are not unlike Asia's *Tetracanthagyna* (p. 120), being enormous and plain brown with dark-rayed wings. Nymphs hide in the crevices of pieces of wood that lie in streams.

AESHNIDAE—UNAFFILIATED EURASIAN AND AMERICAN GENERA

SPECTERS, SPOOKHAWKERS, AND POSSIBLE KIN

Adult *Boyeria* seem allergic to sunlight. The author has watched a Western Specter (*B. irene*) dash about in the shade of a parked car like a caged animal, perhaps mistaking the sunbaked road for water. Normally, males closely track the overhanging banks of streams, rivers, and sometimes lakes. Flying low and cautiously, every dark corner is inspected for females. Their world expands as the sun sinks and they aggregate in open fields by nightfall, flying frenetically in pursuit of prey.

The genus's vast but patchy range (**pink and purple on map**) suggests that much ground was lost with the ice ages. The adults differ strongly in each region, while the nymphs are alike wherever species survive: spinier than other aeshnids with

LEFT | Western Specter (*Boyeria irene*) male in flight.

TOP RIGHT | A male Eastern Specter (*Caliaeschna microstigma*) from Turkey.

MIDDLE RIGHT | Rare photograph of a female Frecklewing (*Allopetalia pustulosa*) ovipositing in the Andes of Ecuador.

LOWER RIGHT | A male *Periaeschna magdalena* patrolling a stream in Vietnam.

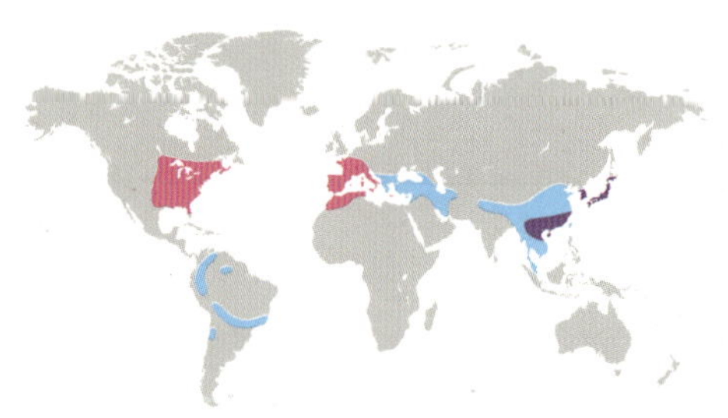

DIVERSITY
93 species of shady (usually forested, often mountainous) streams in Eurasia, and North and South America

TAXONOMY
Affinities of genera *Aeschnophlebia* (32 species, long known as *Planaeschna*, see p. 121), *Allopetalia* (2), *Boyeria* (7), *Caliaeschna* (1), *Cephalaeschna* (30), *Gynacanthaeschna* (1), *Limnetron* (2), *Periaeschna* (12), *Petaliaeschna* (5), and *Racenaeschna* (1) still too uncertain to assign them to proposed subfamilies such as Allopetaliinae, Boyeriinae, and Caliaeschninae

horns on the head's backside. Two North Americans, known as spotted darners, are brown with bright eyelike marks on the thorax sides. Two Mediterranean species don military camouflage blotched dull green and dark gray. Three east Asians recall goldenrings, however: black with yellow banding (p. 86).

Other stream-dwelling aeshnids in southern and eastern Eurasia (**blue on map**) are similarly secretive, concealed in their dappled forest habitat by black bodies with contrasting yellow to green markings. Like *Boyeria*, the nymphs probably cling to rootlets under streambanks, below rocks, or among woody and leafy debris, often in dense forest and mountain areas, sometimes at considerable elevation.

Most seem highly localized and specialized. *Cephalaeschna* nymphs on Taiwan may emerge at night to hunt on land, for example. Unsurprisingly, three-fifths of the species of harvesthawker (*Cephalaeschna*), spookhawker (*Periaeschna*), and phantomhawker (*Aeschnophlebia*) were only named this century; *Gynacanthaeschna* has features intermediate between the first two genera. This elusiveness, plus the species' shade-hugging and dusk-loving habits, inspired their ghostly names. Many fly only in late summer and the fall, hence the name "harvesthawkers." The co-occurring and similarly black-and-yellow *Petaliaeschna* have bright yellow legs and fly in spring. The small Eastern Specter (*Caliaeschna microstigma*), marked intricately with blue, extends from Iran to the southern Balkans.

South America's furtive stream aeshnids (**also blue**) are least known. With its marbled body and dark-flecked wings, the Frecklewing (*Allopetalia pustulosa*) from the equatorial Andes' cloud forest (a related species is confined to Chile) is particularly odd. Forest-darners (*Limnetron*) are restricted to Andean and Atlantic rainforests, while *Racenaeschna* is known only from females collected on Venezuela's tepuis.

AESHNIDAE—GOMPHAESCHNINAE

PYGMY DARNERS, BOGHAWKERS, AND PADDLETAILS

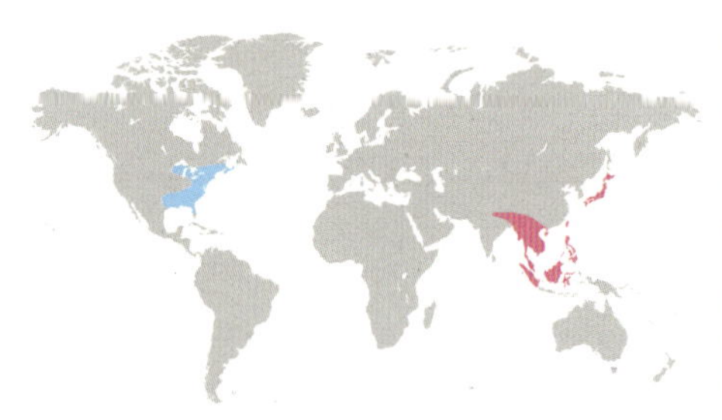

DIVERSITY

42 species of small, peaty wet spots, usually in forest and mostly in South and East Asia, with just two occurring in eastern North America

TAXONOMY

Genera *Gomphaeschna* (2 species; **blue on map**), *Linaeschna* (1), *Oligoaeschna* (19), *Sarasaeschna* (18), and *Sundaeschna* (2)

Pygmy darner (*Gomphaeschna*) females are peculiar for having a terminal club like Gomphidae, the family of clubtails (p. 92). They are named, however, for the male's split and often splayed lower clasper, which is like a two-pronged fork or fishtail, and is characteristic of both its subfamily and that distantly related group.

These slight hawkers, which have notably open wing venation, set off on their separate evolutionary path before the aeshnids in the previous accounts diversified. All but the two *Gomphaeschna* species are from Asia. Their males have distinctly spindle-shaped abdomens, their bases swollen and then sharply waisted, like those of many other aeshnids, but wide in the middle again, narrowing to a slender tip.

All are highly elusive. Over half the boghawkers (*Sarasaeschna*) were only named this century, for example. Multiple species may occur together, but their flight season is short and early, while their remote habitats are scattered across the mountains from the eastern Himalayas to Indochina and Japan. Males are black with bold green-yellow flecks and hang near (or fly over) cool damp areas in forest that often contain peatmoss but hardly any open water. The nymphs cling tightly to the soaked plant matter that fills small depressions in the wet ground. America's *Gomphaeschna* similarly grasps onto the underside of logs in shallow pools carpeted with wet moss. Adults also fly mostly in spring, but are variegated in color and rest on trees to warm up (photo p. 14), thus recalling *Aeshna* species (p. 115).

Except for *Linaeschna polli*, known from just a few specimens from Borneo, most species from tropical Southeast Asia are classified as paddletails (*Oligoaeschna*). They likely favor wet depressions too, probably also temporary ones, in peat forest as well as wooded floodplains and swamps. Adults are most active late in the day. Hunting in clearings at dusk, the females stand out with tinted wings and spatulate appendages, which inspired their common name. Described only in 2018 from lowland rainforest in Myanmar and Vietnam, the genus *Sundaeschna* seems intermediate to *Sarasaeschna* and *Oligoaeschna*.

OPPOSITE | A male of the Japanese Bog-hawker (*Sarasaeschna pryeri*) showing its distinctive abdomen shape and split lower clasper.

RIGHT | Male of the Harlequin Darner (*Gomphaeschna furcillata*), which flies mostly in spring.

FAMILY: AUSTROPETALIIDAE

REDSPOTS

LEFT | A male of the White-dotted Redspot (*Hypopetalia pestilens*) from the Patagonian Andes.

OPPOSITE | A female of the Waterfall Redspot (*Austropetalia patricia*) from southeast Australia.

Until 30 mya, Antarctica had forests. Their dragonflies may have been like those limited to the south's cooler habitats today, such as Neopetaliidae (p. 88), Hemigomphini (p. 105), Petalurinae (p. 110), and Austropetaliidae. Together with Aeshnidae (p. 112), this family branched off the dragonfly tree before all other extant families diverged, so may be considered the evolutionarily most isolated dragons.

The White-dotted Redspot (*Hypopetalia pestilens*) from the Patagonian Andes is the archetype of an archaic dragonfly, with a powerful brown body, small glassy-gray eyes, ghostly white flank dots, and seven red blotches leading each wing. With reduced wing spots and yellow-striped bodies, its smaller compatriots, the striped redspots (*Phyllopetalia*), are barely less odd, having side flaps to the abdomen tip. Southeast Australia's *Austropetalia* species are quite

DIVERSITY
11 species of streams, runnels, seeps, and splash zones in the temperate south of South America, Australia, and Tasmania

ADULT HABIT
Rather large dragonflies with ovipositors (in females) that differ from the similar-sized and shaped aeshnids by their smaller (slightly separated) eyes and 5–7 reddish blotches on the wings' leading edges

TAXONOMY
Genera *Archipetalia* (1 species), *Austropetalia* (3), *Hypopetalia* (1), *Phyllopetalia* (6)

similar, as is the Tasmanian Redspot (*Archipetalia auriculata*). The nymphs have been called the most grotesque in Odonata, being shorter and more strongly built than aeshnids, with gnarly limbs and thick, warty skin. The broad abdomen's armored appearance may recall woodlice. Hiding among pebbles and debris or clinging under logs and rocks, some nymphs inhabit streams a few meters wide, but they occur mainly in tiny and sometimes very diffuse flowing waters, so might be semiterrestrial too. Whether they venture onto land much is unconfirmed, however, as is their possibly six- or seven-year development time.

Although *A. auriculata* sits flat on the ground by streamlets, adults are furtive and seldom settle near water. *H. pestilens* males fly low over deeply shaded streams in mountain forests, but most species may only be seen when hunting along forest edges or when females lay their eggs in moist logs.

While most isolated damsel lineages are tropical (pp. 194–232), eccentric dragons survived in cooler climes (p. 20): Cordulegastridae, Chlorogomphidae, and Epiophlebiidae (pp. 86, 90, and 130) in the north, and various "corduliids" in both hemispheres (pp. 74–9). Many even look quite similar and were therefore thought to be related, as expressed by names with "petal" in them (pp. 88 and 110).

Perhaps more dragonflies survive in colder regions because larger bodies retain heat better. Those would then also be bigger than their tropical relatives, with expanded body parts: narrowly touching eyes, for example, may be linked to many cold-adapted dragons' swollen and wide-faced heads. Their nymphs may need years to grow so large, putting strong evolutionary pressure on them (p. 9). Innovations allowing certain groups to dominate worldwide (such as great size, but perhaps also the "cavilabiate" mask; p. 86) may thus have arisen in cooler settings first, remnants of this history persisting there to this day.

FAMILY: EPIOPHLEBIIDAE
DAMSELDRAGONS

LEFT | The gap in the damseldragons' known range was narrowed substantially with the description of *Epiophlebia sinensis* from China's northeast in 2012 and the subsequent discovery of a new subspecies of *E. laidlawi* in the country's southwest, shown here.

OPPOSITE | Like all damselflies but only three dragonfly families (pp. 110–129), damseldragons such as *Epiophlebia superstes* from Japan have ovipositors to lay their eggs in plant tissue.

Odonates suit our binary way of thinking, with almost equal numbers of species classified as dragonflies and damselflies (p. 14). Just a single surviving genus, *Epiophlebia*, challenges that convenient subdivision. It is quite appropriate, therefore, that scholars have argued about which suborder this genus belongs to ever since the Austrian entomologist Anton Handlirsch coined the (internally contradicting!) amalgam "Anisozygoptera" in 1906.

While quite small, the yellow-spotted black adults look like typical "primitive" dragonflies, recalling austropetaliids most with their wide faces and ovipositors (p. 128). The coloration and spread eyes also bring cordulegastrids

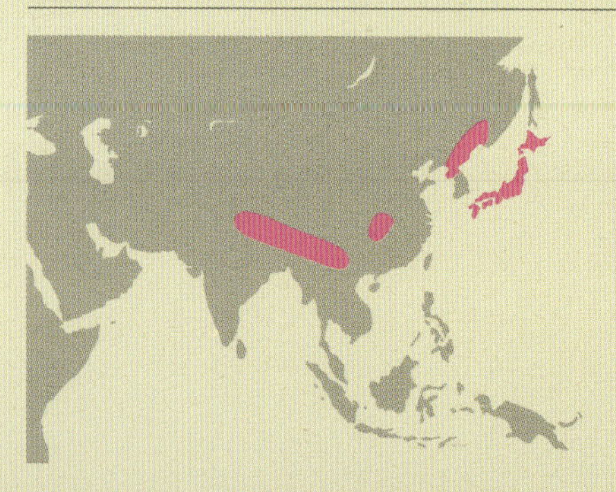

DIVERSITY
3 species of cold forest streams from Japan to the Himalayas

ADULT HABIT
Small (1¾–2 in/4.5–5 cm long) with yellow-spotted black bodies, wide faces, spread eyes, and unique narrow damselfly-like wings that are closed over the hanging abdomen at rest

TAXONOMY
Genus *Epiophlebia* (3 species)

(p. 86) and gomphids to mind (p. 92). The wings are narrow, however, and virtually alike, closed over the abdomen as damselflies do while they hang in the vegetation like aeshnids (p. 112). Just as dragonflies do, male damseldragons use their lower clasper (the so-called epiproct) to grasp the female's head (p. 17). Some scholars have therefore proposed to expand Anisoptera to include *Epiophlebia* and rename the suborder Epiprocta.

Only the Japanese (*E. superstes*) and Himalayan Damseldragons (*E. laidlawi*) were known until 2012, when the Manchurian Damseldragon (*E. sinensis*) was described from northeast China and adjacent North Korea. Adults are inconspicuous and only present for a month or so in late spring, moving about their cold forest streams with a rapid but comparatively uncoordinated flight.

For most of their five- to eight-year development, the nymphs cling to rocks in the current, but the final few months appear to be spent crawling on land nearby. While looking like a cross between a gomphid and an aeshnid, they produce a grating noise when disturbed by rubbing their hind legs against rough patches on the abdomen; they are the only odonates (in any life-stage) equipped to make sound!

Especially in the Jurassic, many more groups with such weirdly mixed characters appear to have existed (p. 15). Damseldragons could be the sole survivors of a lineage that is over 180 million years old, therefore. The extant species are genetically so close, however, that they may only have become isolated in their current mountainous haunts when the last ice age ended. Just 20,000 years ago, therefore, these "living fossils" probably moved around Asia as freely as many pond species do today!

Molecular studies also revealed that, even after the two suborders diverged, genes coding for damselfly-like characters slipped into dragonfly-like species' genomes due to interbreeding. That was still many millions of years ago, but the damseldragons' hybrid look may thus not stem from before all other "mixed" odonates died out. How every odonate group is related and should be named, and which deserve the rank of suborder, might be debated forever, therefore.

ZYGOPTERA

FAMILY: COENAGRIONIDAE

POND-DAMSELS

LEFT | Nothing might demonstrate the odonates' evolutionary versatility more than the 1 in (2.5 cm) long Sedgling (*Nehalennia speciosa*, p. 164), which is related more closely to rainforest giants (up to 6 in/16 cm long!) in tropical America (p. 156) than to the *Coenagrion* bluets flying nearby in its northern Eurasian bog (p. 148).

OPPOSITE TOP | Nymph of Europe's Crescent Bluet (*Coenagrion lunulatum*) showing two distinct halves in its leaflike gills.

OPPOSITE | Although more diversely colored than any other damselfly family, the typical coenagrionid is a small blue species such as Africa's Tiny Bluet (*Azuragrion vansomereni*) (pp. 134 and 148).

DIVERSITY
1,396 species at all types of fresh (especially standing; sometimes brackish) water almost anywhere odonates occur

ADULT HABIT
Extremely varied, from the tiniest to the longest odonates. Especially at standing waters, however, most smaller damselflies with a simple square mesh of veins in narrow wings that are shut at rest belong to this family

TAXONOMY
At least 6 subfamilies with numerous tribes are recognized, of which Ischnurinae and Ischnurini have most species, although the limits of some of these subgroups are still uncertain, notably those with a (mostly) ridged face (pp. 152 and 156)

As with Libellulidae in dragonflies (p. 34), the first damselfly most people encounter is likely one of some 1,400 species in 120 genera placed here, in the largest odonate family by far. With the stream-damsels (p. 166) and wiretails (p. 178) they also form the order's greatest radiation: as most new species are discovered in the superfamily Coenagrionoidea too, well over 2,000 must exist, a third of Odonata (and two-thirds of Zygoptera)!

Unsurprisingly, the species are enormously varied. While most dragonflies are bulkier, Coenagrionidae contains the longest-bodied and widest-winged odonates. Most species are small, however, so (equally unsurprisingly) the tiniest odonates are coenagrionids too.

As the name "pond-damsels" suggests, coenagrionids are like libellulids in handling the vagaries of standing waters best. Just as for those "pond-dragons," their success is due particularly to their ability to handle change well in general (p. 136), with hundreds of species actually confined to rivers and streams (pp. 150–5). Dispersal is probably the

key: while small, many species reach far-off places easily, likely carried by air currents (p. 12).

The simple design of the similar-looking coenagrionid nymphs' leaflike terminal gills may provide the evolutionary flexibility needed to adapt to the diverse places their adults reach: a hinge-like suture at midpoint allows the terminal halves to flex and develop quite differently than the basal halves. They can be larger to respire better in a fetid pool, or longer for more thrust, as provided by a tailfin, in fish-infested waters.

COENAGRIONIDAE—ISCHNURINAE—*ENALLAGMA*

AMERICAN BLUETS

Just like their dragonfly counterparts in Libellulidae (p. 34), coenagrionids are incredibly diverse in color. The archetypical species, nonetheless, are small, extensively pale blue damselflies marked finely with black. Such "bluets" are found all over the family tree (p. 148), however, and even beyond (p. 242).

Indeed, when coining the name *Enallagma* (Greek for "interchange") two centuries ago, Toussaint de Charpentier already advised close scrutiny to separate these confusingly similar damselflies! Half of the species that still carry this name are classic bluets, so alike that their identification literally gives us the blues. They only evolved across North America in the past few hundred millennia, as ice ages came and went, even slipping across to Eurasia where a few virtually identical species remain.

While some of the remaining American bluets are blue too, most of the species are purple, red, orange, yellow, or, like the Rainbow Bluet (*E. antennatum*), multicolored. Having survived in warmer regions, with the Neotropical Bluet (*E. novaehispaniae*) even extending far into South America, they evolved much earlier than their consistently blue relatives. Genetics suggest that California's Exclamation Damsel (*Zoniagrion exclamationis*) may fall with them, as may the Caribbean Damsel (*Enacantha caribbea*) from Cuba and the Middle American mainland nearby.

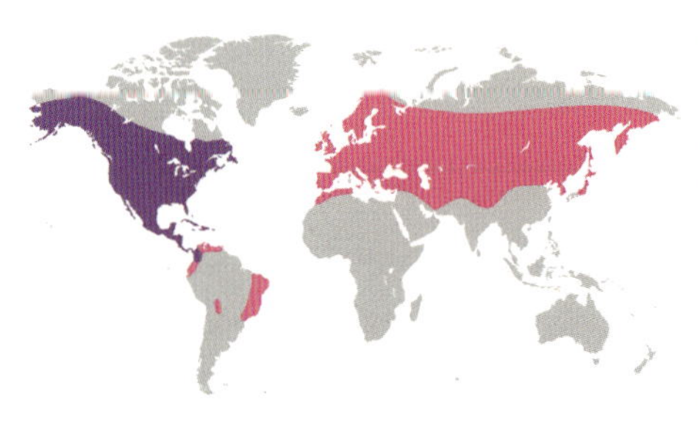

DIVERSITY
45 species at broad range of standing and flowing waters; only a few (**pink on map**) extend beyond North and Middle America to South America, Eurasia, and northern Africa

TAXONOMY
Genus *Enallagma* (43 species) with subgenera *Enallagma* (24) and *Chromatallagma* (19) into which the genera *Enacantha* (1) and *Zoniagrion* (1) may fall too

TOP | Occurring widely across temperate Eurasia, the Common Bluet (*Enallagma cyathigerum*) is one of the few species in the genus that extended beyond North America.

LOWER | The more colorful North American *Enallagma* species, such as the Rainbow Bluet (*E. antennatum*), are separated in the subgenus *Chromatallagma*.

FORKTAILS AND BLUETAILS

How much control does a damsel have once a male has convinced her to mate? Only in 2022 researchers revealed that females can prick the male's sperm pouch with a small spine at their ovipositor's base. Although some species lack this vulvar spine, its presence had long been known as the defining feature of *Enallagma* (see opposite) and its many relatives.

With over 400 species, they make up the largest coenagrionid subfamily, Ischnurinae. Encompassing at least 70 species that between them have reached every habitable corner, *Ischnura* is its most familiar genus. Most *Ischnura* males have two small prongs on the abdomen tip and a blue tail-light, inspiring the names forktail (Americas) and bluetail (elsewhere), although other coenagrionids have

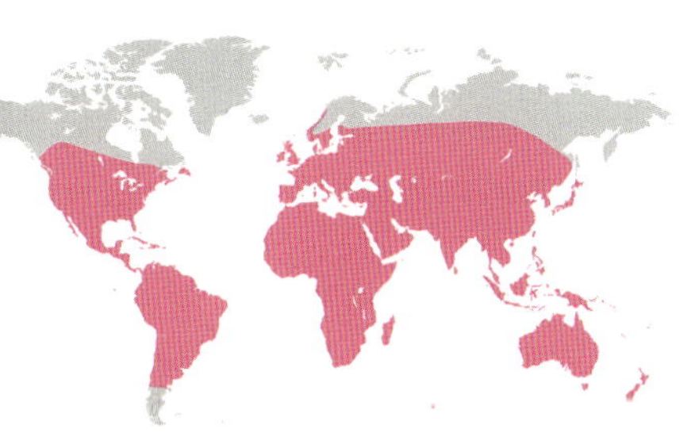

DIVERSITY
78 species, which potentially occur almost anywhere where odonates are present

TAXONOMY
Genus *Ischnura* (70 species), as currently delimited, may include the Pacific genera *Amorphostigma* (3), *Hivaagrion* (2), and *Pacificagrion* (2) too

ABOVE | A mating pair of Common Bluetails (*Ischnura elegans*) in the UK, with the female's vulvar spine buried into the male's second abdomen segment, where his secondary genitalia are located (p. 17).

these features too. Males also frequently have oddly colored (often two-toned) or proportioned stigmas. While ischnurines generally owe their success to opportunism, no odonates mastered this better than *Ischnura*, which cope well in habitats that are too unstable, eutrophic, brackish, hot, cold, or remote for their competitors. As in the sister-genus *Enallagma*, the many north-temperate species probably arose in response to new habitats opening up after every ice age, especially in North America.

In the tropics, *Ischnura* is represented best on islands and (the equally insular) mountains. Species that evolved in isolated forest streams, in particular, may look unusual and were thus placed in separate genera: Samoa's *Amorphostigma* and *Pacificagrion*, and the Marquesas Islands' *Hivaagrion*. Slightly more "normal" bluetails such as India's *Rhodischnura*, the Bonin Islands' *Boninagrion*, and *Oreagrion* in New Guinea's highlands have already been reclassified as *Ischnura*.

The few species of tropical lowlands are adaptable wanderers, tolerating urban waters. Africa and Asia's Tropical Bluetail (*I. senegalensis*) even breeds in greenhouses, and is the most-exported odonate (p. 46). Those that are widespread in the American tropics did not originate in nearby North America but across the ocean.

Possibly because there might be no males where they go next, Wandering Bluetails (*I. aurora*) in Oceania and *I. praematura*, at high elevations in China, are the only odonates that may mate as

TOP LEFT | A mature female of the Tropical Bluetail (*Ischnura senegalensis*)—widespread in Africa, the Middle East, and southern and eastern Asia—colored like a male.

LOWER LEFT | Another female of the Tropical Bluetail (*Ischnura senegalensis*) from Kerala, India, with bright orange coloration.

TOP RIGHT | Despite lacking a distinctive blue tail-light, India's cute Pixie Bluetail (*I. nursei*), once placed in its own genus, *Rhodischnura*, was confirmed by genetics to belong to *Ischnura*.

LOWER RIGHT | This male Madagascar Bluetail (*Ischnura filosa*), covered with parasitic water mites, represents the largest group within the family, the wildly successful Ischnurinae.

soon as the females emerge from their nymphal skin. As their ecological strategy often allows bluetails to occur at high densities, competing males may also simply be trying to get in first.

Indeed, competition between the sexes can be intense too. Besides having camouflage colors, females of half of the *Ischnura* species (and many other coenagrionids) have male-like forms, which may suffer less unwanted attention. Many fresh females are luminous orange, moreover, perhaps to signal they are still eggless and best left alone. Some species even have three female morphs, which change color with age.

Unlike most coenagrionids, females oviposit alone, preferring the morning or evening, and may attack approaching males. Rivals can remove their sperm while the females wait to oviposit, so males may force females to remain in copulation for hours. It is then that pricking him in the genitalia comes in handy. Indeed, his sperm pouch may be appropriately padded!

After reaching the Azores from North America, the Citrine Forktail (*I. hastata*) somehow did away with the opposite sex altogether: new generations of females come forth from unfertilized eggs. While common in other insects, this is the only known case of parthenogenesis in odonates.

COENAGRIONIDAE—ISCHNURINAE—*NESOBASIS* AND RELATED GENERA

ISLAND-DAMSELS

BELOW | Members of this group have fine legs, none more so than Australia's tiny spindlylegs (*Austrocnemis*) that settle on floating vegetation, such as this male Splendid Spindlyleg (*A. splendida*).

OPPOSITE TOP | Fiji's Red-eyed Island-damsel (*Nesobasis erythrops*) is among the most spectacular of almost 40 closely related species found on forest streams there.

Early this century, curious researchers ventured to Fiji. Not only do its many *Nesobasis* and *Nikoulabasis* species occur nowhere else, but some have more colorful females than males. Might sex roles be reversed? Males of other species had never (or only rarely) even been seen, while females were common. Could they breed asexually, like *Ischnura hastata* on the Azores (p. 137)?

DIVERSITY

52 species at all freshwaters in the Fijian and Vanuatuan archipelagoes, with a few in Australia and New Guinea too

TAXONOMY

Genera *Austrocnemis* (3 species), *Nesobasis* (28), *Nikoulabasis* (10), *Thaumatagrion* (1), and *Vanuatubasis* (10)

RIGHT | While the species of *Nesobasis* from Fiji and *Vanuatubasis* from Vanuatu are closely related, some Fijian species such as this male *N. comosa* are more distant, so the genus *Nikoulabasis* was erected in 2023 to incorporate them.

Two decades later, the mystery has only grown. The number of species has since doubled to almost 40. These differ widely in color, shape, and size, but little in ecological preferences, with up to 12 living on the same forest stream. Females do not defend territories, nor procreate by themselves. Males remain rare in three species, though, and are still unknown in two.

And while genetics show that the damsels dispersed between (and became isolated on) the archipelago's islands many times, this cannot fully explain how so many (and such varied) species could arise and coexist, nor why some females are so conspicuous. One dispersal event went 500 miles (800 km) west to Vanuatu: every *Vanuatubasis* species is confined to a single island, but eight were only named in 2022. Just six of the 13 main islands have been studied so far, moreover.

Two related genera in Australia and New Guinea suggest that the Melanesian radiation originated there, as they are similar, with fine legs and dark bronzy bodies. The tiny spindlylegs (*Austrocnemis*), with their unusually long legs, rest on lilypads and other pond vegetation, while the black-winged *Thaumatagrion funereum* lives in swamp forest.

COENAGRIONIDAE—ISCHNURINAE—*MEGALAGRION*

PINAPINAOS OR HAWAIIAN DAMSELS

Hawaii's damselflies may form Odonata's most spectacular radiation. Not only are most *Megalagrion* adults big and colorful, marked boldly with black and red, they occupy the widest habitat range of any genus, benefitting from all possible freshwater on the islands. Some even breed away from water!

This is all the more remarkable because (as evolutionary studies show) these damsels moved more freely between islands than habitats: a stream species, for example, can cross 30 miles (50 km) of ocean to another stream more easily than adapt to an adjacent pond (p. 12). Only the seven most distinct species still occupy the suspected ancestral habitat: lowland pools. Of the others, two similar species breed only in water held in leaf axils, and three in fast-flowing streams.

The 11 closest relatives diversified across niches available only on such rugged and rainy islands, such as dripping rockfaces, mossy seeps, and upland bogs. Nymphs of some even survive among damp leaf-litter under dense ferns, their terminal gills being strongly reduced. The species' diversification was probably facilitated by the absence of terrestrial predators such as ants and mammals, and aquatic insects like mayflies, stoneflies, and caddisflies. Today a third is threatened with extinction (photo p. 29).

DIVERSITY
23 species at almost any natural wet habitat in the Hawaiian archipelago

TAXONOMY
Genus *Megalagrion* (23 species)

ABOVE | Scarlet Pinapinao (*Megalagrion vagabundum*) pair mating by a stream on Kauai. These Hawaiian damsels have been known by long and varied names, such as Scarlet Kauai Damselfly, Hawaiian Upland Damselfly, and Flying Earwig Hawaiian Damselfly. For the group, it may be appropriate to adopt the local word for damselfly, Pinapinao.

PALEOTROPICAL BLUETS AND SLIMS

Being partially replaced by its nearest relative, Agriocnemidinae (p. 146), Ischnurinae is represented poorly in the Old World tropics. Scattered widely across Africa, Asia, and Australia, most species are very slender and are currently classified as slims (*Aciagrion*). The dull-brown adults may shelter in grassy undergrowth for months, only turning blue as pools fill up with rain.

Lancets (*Xiphiagrion*) favor barren ponds in the archipelagos between Asia and Australia, while *Austroallagma* is confined to the Lesser Sundas.

African bluets (*Africallagma*, *Pinheyagrion*, *Proischnura*) are largely restricted to highland pools and marsh in Africa and Madagascar, while sailing bluets (*Azuragrion*) also extend to the Comoros and Arabia (see photo p. 133). The latter's tiny species settle on the surface tension of puddles and drift with the wind. Until 2002, these bluets were erroneously placed in *Enallagma* (p. 134), but further name changes in this group are likely. *Azuragrion* is close to South Asia's Little Bluet (*Amphiallagma parvum*), for example.

Like other ischnurines this group colonized islands too, with larger and more varied adults. The stunning Socotra Bluet (*Azuragrion granti*; photo p. 30) on the arid island of Socotra, for example, has almost double the wingspan of its relatives. Most colorful are the bluetips (*Coenagriocnemis*), which inhabit rocky forest streams and rivers on the Mascarenes. Males of the Orange-legged Bluetip (*C. rufipes*) from Mauritius show their orange faces and legs in aggressive displays.

ABOVE | This male Australian Blue Slim (*Aciagrion fragile*) is one of many very similar species in the group.

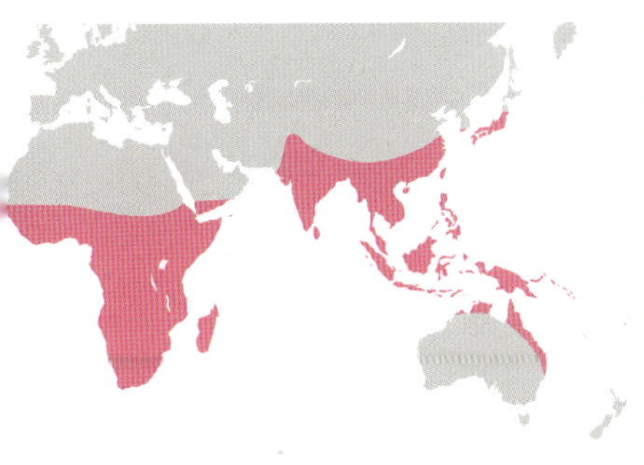

DIVERSITY
57 species of mostly standing waters in warmer parts, from Africa to northern Australia

TAXONOMY
Genera *Aciagrion* (27 species), *Africallagma* (12), *Amphiallagma* (1), *Austroallagma* (1), *Azuragrion* (6), *Coenagriocnemis* (4), *Pinheyagrion* (1), *Proischnura* (3), and *Xiphiagrion* (2)

WEDGETAILS, CORALS, AND KIN

While the ischnurines in the previous accounts were conquering the rest of the world, a separate group of them (often also with the female's diagnostic vulvar spine; p. 135) diversified in the American tropics. Although some are unusual among ischnurines for radiating mostly within forested lowlands (next profile), most of these "leptobasines" favored more open landscapes too.

Over half of the latter group's species are classified either as wedgetails (*Acanthagrion*) or corals (*Oxyagrion*).

TOP LEFT | Found in *Andinagrion*, *Oxyallagma*, and *Protallagma* too, the striking color combination of Brazil's Blue-tipped Coral (*Oxyagrion simile*) is notably common in this group.

LOWER LEFT | Inhabiting sedgy bog and marsh across North America, the two near-identical species of minute and stocky American red damsels (*Amphiagrion*) may be related most closely to this group or that treated next. Here an Eastern Red Damsel (*A. saucium*) is shown.

OPPOSITE | A male Pacific Wedgetail (*Acanthagrion trilobatum*) in Panama demonstrating its wedge-like "tail."

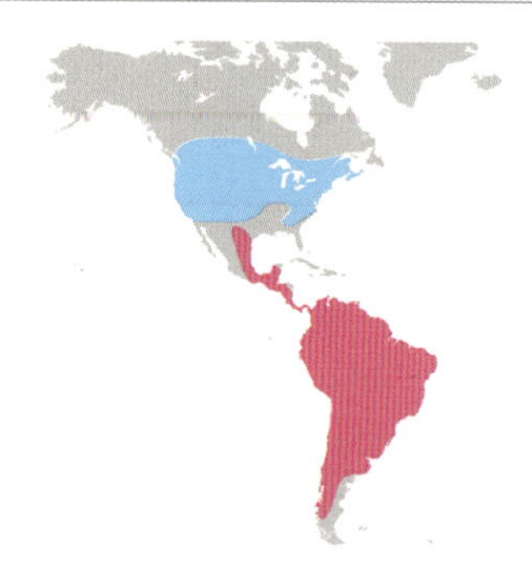

DIVERSITY

119 species at all kinds of (often sunny) freshwaters from the warmer parts of North America to the coldest in South America

TAXONOMY

Possibly synonymous genera *Acanthagrion* (42 species) and *Oxyagrion* (26) plus closely related (also partly synonymous?) *Acanthallagma* (3), *Andinagrion* (3), *Argentagrion* (2), *Cyanallagma* (8), *Fluminagrion* (1), *Franciscagrion* (2), *Franciscobasis* (1), *Homeoura* (4), *Mesamphiagrion* (14), *Negragrion* (1), *Oreiallagma* (5), *Oxyallagma* (2), *Protallagma* (2), and *Tigriagrion* (1); genus *Amphiagrion* (2; **blue on map**) may be near this group or that on p. 144

Translating from the Greek for "thorn -damsel" and "sharp-damsel" (p. 148), respectively, their names refer to the (pointed) hump and sloping claspers on the male's abdomen tip, which form a wedge. Both seem as interchangeable in use as in meaning, though: while most male *Oxyagrion* are largely red (compare p. 163) and many *Acanthagrion* are typical bluets, marked black and blue (p. 148), other species are colored rather differently, and the two genera may ultimately prove not to be separable.

Coinciding with this diversity in color, the 68 species occupy a wide array of standing and flowing waters, typically with vegetation like waterweeds and bankside grass, from Texas to Argentina. Smaller genera with similar male structures but unusual colors, like Brazil's orange-and-black striped Tiger Damsel (*Tigriagrion aurantinigrum*), probably form part of this complexity. Maroonwing (*Acanthallagma*) males are small but have broadened wings, their basal halves dark amber. Perching at sunny streams in Amazonia, they ascend with a jerking butterfly- or bee-like flight when disturbed.

Most smaller genera are still poorly known, with several like *Fluminagrion* and *Negragrion* (both very close to *Acanthagrion*) only named recently. While the marshdamsels (*Homeoura*) occur across tropical South America, many are restricted to the continent's higher and cooler parts, such as *Andinagrion*, *Argentagrion*, *Cyanallagma*, *Franciscagrion*, and *Franciscobasis*, which are found from Patagonia to southern Brazil.

Genera in the tropical High Andes, such as *Mesamphiagrion* and *Oxyallagma*, seem particularly specialized. At least one mountain-damsel (*Oreiallagma*) breeds in water-filled bromeliads in cloud forest, up to 7,500 ft (2,300 m) above sea level. The small and stocky *Protallagma* of the altiplano (mean elevation 12,300 ft/3,750 m) are among the highest-ranging odonates. Often the only species present, they may occur in great numbers.

COENAGRIONIDAE—ISCHNURINAE—*LEPTOBASIS* AND RELATED GENERA

AMERICAN SWAMPDAMSELS AND KIN

Many coenagrionids change color dramatically with age. Both sexes of *Leptobasis* and its nearest relatives, for example, may be wholly orange at emergence before dark markings come through gradually, changing from orange to green and then black. Pale markings may go from orange through green to blue. Consequently, the thorax can be striped stunningly with blue on orange, or blue on green. Some species become largely dark, but many retain a glowing yellow to red abdomen tip.

As in *Ischnura* (p. 136), the males' forewing stigmas may be modified—for example, being black with glistening white rims in pearlwings (*Anisagrion*). The wingtips can also be darkened, either with pigment or densely reticulated veins. Pearlwings and the related Black-and-white Damsel (*Apanisagrion lais*) and Painted Damsel (*Hesperagrion heterodoxum*) favor rank vegetation of still or sluggish waters (river backwaters, stream pools, grassy seeps) from the southwestern USA into the northern Andes.

DIVERSITY

46 species at a wide range of swampy habitats, mostly in the American tropics

TAXONOMY

Genera *Anisagrion* (4 species), *Apanisagrion* (1), *Calvertagrion* (5), *Denticulobasis* (3), *Dolonagrion* (1), *Hesperagrion* (1), *Leptobasis* (9), *Leucobasis* (1), *Mesoleptobasis* (5), *Telagrion* (1), *Tuberculobasis* (13), *Tukanobasis* (2); genus *Amphiagrion* (2) may be near this or previous group (p. 142); presumed protoneurines *Junix* (1) and *Proneura* (1) may be related too (p. 155)

The more slender American swampdamsels (*Leptobasis*) found there and across the Caribbean inhabit seasonally flooded lowlands, mostly with forest. The bluelines or longtails (*Denticulobasis*, *Mesoleptobasis*, *Tuberculobasis*, and *Tukanobasis*), Varzeadamsel (*Dolonagrion fulvellum*), and Candescent Swampdamsel (*Leucobasis candicans*) prefer similar shady habitats in Amazonia, as does the Flametip (*Telagrion longum*) in Brazil's Atlantic Rainforest. With their very long and often reddish-tipped abdomens, they recall distant relatives such as *Metaleptobasis* (p. 161) and *Psaironeura* (p. 154) found alongside them.

At ¾ in (2 cm) long, the Amazonian *Calvertagrion* species are among the smallest damsels. Hiding in vegetation, these color-morphing midgets recall *Agriocnemis*, some similarly become black covered in thick white pruinosity. This reflective shield might protect such small damselflies from dehydration, but also makes males stand out in the shade. Indeed, while *Leptobasis* and its allies are often longer (probably linked to hovering, see p. 155), they are the American equivalents of Agriocnemidinae in many ways (next profile).

OPPOSITE | The gaudy male of the Painted Damsel (*Hesperagrion heterodoxum*) can only be found in Mexico and the adjacent USA.

ABOVE | The Amazonian genus *Calvertagrion*, including this male *C. minutissimum* from Brazil, strongly recalls the Old World wisps.

RIGHT | This male Red-tipped Swampdamsel (*Leptobasis vacillans*) from Mexico is quite mature, but may become blacker still.

COENAGRIONIDAE—AGRIOCNEMIDINAE

WISPS, MIDGETS, AND SHADEFLIES

While small to us, odonates are big by insect standards, probably because the physical demands of their flying abilities do not allow them to be smaller. With some approaching ⅔ in (17.5 mm), while being very slim, wisps (*Agriocnemis*) must include the tiniest species (compare p. 49). Sweeping a net through the dense grasses among which they often live may be the best way to find them. Species of coarser vegetation, such as swamp forest, are generally larger. Africa's Papyrus Wisp (*A. palaeforma*) is over 1 in (2.8 cm) long, for example, and found only among those giant sedges.

Over 40 species currently placed in *Agriocnemis* make up the bulk of the subfamily, which recalls its apparent sister-group Ischnurinae in many respects (p. 136). The Little (*A. exilis*) and Wandering (*A. pygmaea*) Wisps pop up at almost any grassy puddle in tropical Africa and Asia, respectively, and range across the Indian and Pacific Ocean archipelagoes too.

As in the previous group (p. 144), both sexes show dramatic color changes. Found across the subfamily's range, typical *Agriocnemis* emerge largely red but blacken with age, leaving only the tail-end orange in males. Frequently, even that darkens and dense pruinosity develops (compare p. 145). Many Asian species placed in *Agriocnemis* are more varied, however, with stunningly white, pale blue, or yellow mature males, marked finely with black. These may be closer to the subfamily's 18 remaining species.

Placed in two poorly defined genera, the latter are longer and found mainly in forest swamp habitats in tropical Asia and Australasia. At least five midget (*Mortonagrion*) species prefer coastal habitats such as mangroves. Common at shaded waters from southern Asia to northern Australia, the Variable Shadefly (*Argiocnemis rubescens*) may be largest (though still modestly sized) and most varied species of all. Starting off resembling a *Ceriagrion* species (p. 163)

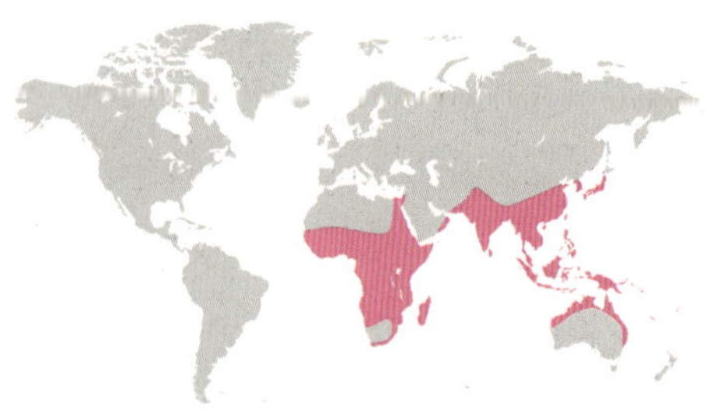

DIVERSITY
62 species at a wide range of still or slow-flowing waters (typically well-vegetated, often temporary) from the tropics of Africa to Australia

TAXONOMY
Poorly delimited genera *Agriocnemis* (44 species), *Argiocnemis* (2), and *Mortonagrion* (16)

ABOVE LEFT | Africa's Papyrus Wisp (*Agriocnemis palaeforma*) lives among the giant sedge for which it is named.

ABOVE RIGHT | The Asian *Agriocnemis* species, such as these mating White Wisps (*A. pieris*), are more varied in appearance than those in Africa, suggesting some may belong to unrecognized genera.

by being entirely red with a yellow thorax and head, it ends more like some *Pseudagrion* (p. 150), black marked with blue. Its genus name with the "r" and "g" inverted relative to *Agriocnemis* is almost as confusing!

RIGHT | A male Variable Shadefly (*Argiocnemis rubescens*) from Singapore transitioning from largely red to entirely black-and-blue.

COENAGRIONIDAE—COENAGRIONINAE

EURASIAN BLUETS AND POSSIBLE KIN

LEFT | *Coenagrion* males typically guard the egg-laying female in an upright position, as in these European Azure Bluets (*Coenagrion puella*).

RIGHT | The Red-and-Blue Damsel (*Xanthagrion erythroneurum*) is among Australia's commonest odonates, inhabiting most open waters there.

OPPOSITE LOWER | Although Australia's Swamp Bluet (*Austrocoenagrion lyelli*) was long placed in *Coenagrion*, the male does not appear to "stand up" when accompanying the female in oviposition.

Why are so many damselflies blue? The males' brightly colored bodies might serve to recognize each other, thus avoiding costly failed mating attempts. Open waters, which many coenagrionids prefer, are saturated with light from the sky above, moreover: blue males may still be conspicuous there up-close, to other damselflies, but harder to see from farther away, such as by predators.

Why, furthermore, do so many have *agrion* in their names? Translating from Greek as "living in the wild," that name was first applied to damselflies in 1775, with countless similar names introduced subsequently. For over a century, *Agrion* referred mostly to the only small damsels inhabiting temperate Eurasia's full range of freshwaters, from Mediterranean streams to lakes in mountains and tundra. Confusion about their name's correct use, however, led to its replacement in 1890 with *Coenagrion*, meaning "common-damsels."

Despite providing the family name for all those other common damselflies out there, these ordinary

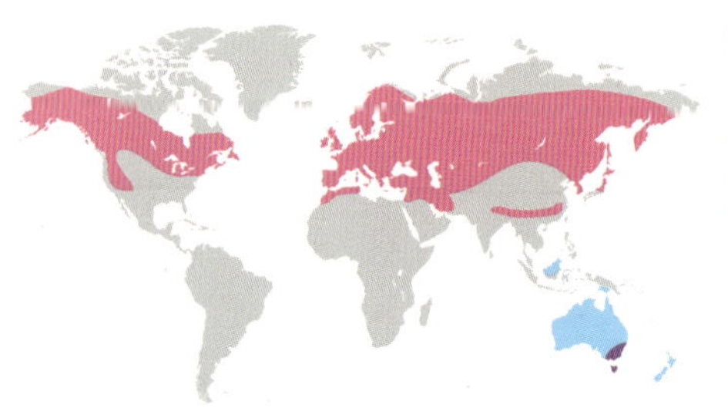

DIVERSITY

27 species at a wide range of standing and flowing waters in temperate parts of the Northern Hemisphere; 12 mostly in similar habitats from New Guinea to New Zealand

TAXONOMY

Genus *Coenagrion* (28 species; **pink on map**) included with certainty; *Austroagrion* (5), *Austrocoenagrion* (1; **purple**), *Caliagrion* (1), *Stenagrion* (2), *Xanthagrion* (1), and *Xanthocnemis* (2) may be nearer *Pseudagrion* (next profile)

bluets' nearest relatives remain unclear. As in *Enallagma*, their ecological counterparts in the New World (p. 134), cold-adapted members of *Coenagrion* managed to switch continents, three species occurring in boreal North America. Australia's Swamp Bluet (*Austrocoenagrion lyelli*) is so similar to that genus in appearance and habitat preference, moreover, that you'd think it was introduced by European settlers! Genetics suggest it is nearer some Australasian genera, however, and the shortlegs (*Stenagrion*) on Borneo and Palawan (**all blue on map**). Most unusually for the family, the latter live in the spray of rainforest waterfalls.

Occupying most open waters, the Red-and-Blue Damsel (*Xanthagrion erythroneurum*) is among Australia's commonest odonates, reaching New Caledonia and Fiji too. It is typically accompanied by one of the billabongflies (*Austroagrion*) in Australia (found also on New Guinea), but replaced by the two redcoats (*Xanthocnemis*) in New Zealand.

Southeast Australia's Large Riverdamsel (*Caliagrion billinghursti*), finally, is like a hefty *Pseudagrion*. Indeed, all these Australasian taxa may prove to belong with Pseudagrioninae (next profile). Even *Coenagrion* might represent a precursor to that subfamily's dominant radiation in the Old World tropics, surviving in the northern periphery just as forementioned genera do in the south (compare pp. 61–4).

COENAGRIONIDAE—PSEUDAGRIONINAE

RIVER SPRITES, BRIGHTEYES, AND KIN

The river sprites (*Pseudagrion*) currently form the largest odonate genus, rivaled only by the distantly related *Argia* in the New World (next profile). Multiple species occur on almost any stream or river, forested or not, from tropical Oceania (where they are called riverdamsels) to Africa. Apparently constrained only by insufficient oxygen and temperature, many species occur on standing waters too, particularly larger ones with enough sun and vegetation.

Males perch on bankside vegetation, or on emerging or floating plants in open water, often close to the surface. Many make lengthy skimming patrols, frequently far from the banks. They often have contrasting yellow to red faces, likely serving as signals that can be seen from a distance (see photo p. 19). Overall, almost any color-scheme realized in damselflies seems possible.

With so many species grouped under one name, a split across three to six genera is anticipated. In Africa, 50 species of generally cooler (running, shaded, or elevated) habitats are structurally distinct from 28 favoring more often stagnant, open, or temporary sites; 31 species confined to Madagascar form a third potential genus. Elucidating the relationships of the 48 species from Asia to the Pacific must unlock the puzzle, many being closer to those found more in forest there, currently placed in *Archibasis*.

The delimitation of this subfamily from Coenagrioninae is similarly unresolved (previous profile). Temperate Eurasia's brighteyes (*Erythromma*) and East Asia's lilysquatters (*Paracercion*) are clearly near *Pseudagrion*, however, and rest similarly on surface vegetation. Perhaps all 12 species are best united under *Erythromma*, which means "red-eye" in Greek, although only two actually have red eyes. Europe's Blue-eye (*E. lindenii*), long placed in *Cercion*, has, nonetheless, already made the move!

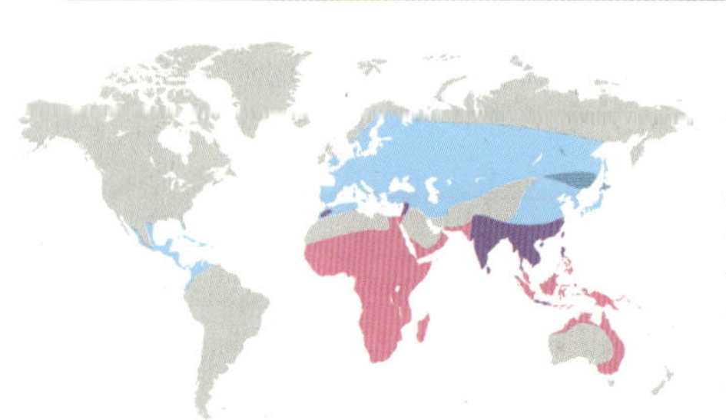

DIVERSITY
179 species at a wide range of warmer flowing and standing waters, mostly from Africa to Australia

TAXONOMY
Genus *Archibasis* (9 species) falls within *Pseudagrion* (156) as currently delimited (**pink on map**); while *Erythromma* (3), *Neoerythromma* (2), and *Paracercion* (9) are clearly close (**blue**), most or all genera in previous profile may be related too

ABOVE LEFT | The male Sulawesi Sprite (*Pseudagrion crocops*) is an especially gaudy representative of its hugely diverse genus.

The two yellowfaces (*Neoerythromma*), found on opposite coasts of Middle America, are the only pseudagrionines to reach the New World. With their bright faces, large and structurally similar male claspers, and habit of perching on exposed vegetation in open water, they fit right in with their relatives from across the ocean!

BELOW | A mating pair of Large Redeyes (*Erythromma najas*) in Finland.

BOTTOM | Restricted to the shores of a freshwater sea in central Africa, the Tanganyika Sprite (*Pseudagrion tanganyicum*) probably evolved from a bright red species found at swamps and ponds nearby (p. 13).

COENAGRIONIDAE—ARGIINAE

DANCERS

Dancers (*Argia*) approach *Pseudagrion* (previous profile) as the largest odonate genus. They are more homogeneous and less likely to be subdivided, however, with more species still expected to be described. Breeding mainly in flowing waters, it is nonetheless tempting to see them as the American answer to *Pseudagrion*.

While most stream-dwelling pond-damsels (also including *Acanthagrion*, for example, p. 142) are associated with aquatic and riparian vegetation, however, *Argia* nymphs live more under stones or in sediment. The rather robust adults like to perch on bare surfaces such as rocks, logs, or the ground. They hold the abdomen rather level or even raised and the wings well above it, rather than slumping down. This posture and their longer leg spines seem linked to a habit of plucking prey from the air, rather than off vegetation. Their bounding flight (hence the common name) is more forceful too, rather than hovering or cruising smoothly.

DIVERSITY
145 species found throughout the (warmer) Americas at most running waters, from tiny trickles and marshy seeps to wide rivers, but sometimes at ponds and lakes too

TAXONOMY
Genera *Argia* (145 species)

TOP LEFT | Most *Argia* males are black and blue, but some such as these Fiery-eyed Dancers (*A. oenea*) from Panama stand out with their cherry eyes and copper thorax.

LOWER LEFT | Males of the Blue-fronted Dancer (*A. apicalis*) need their bright colors to communicate, but turn dark and dull during copulation, probably to reduce the predation risk for them and their mates.

The almost 700 coenagrionid species treated so far are rather closely related and quite similar. All share bright (usually rounded) markings on the back of the head beside the eyes, for example. This suggests that their stream-dwellers evolved from pond-dwellers: adaptations for lush standing waters allowed them to dominate lush running ones too (p. 12). Most of the over 550 treated next, however, lack these postocular spots, while four-fifths of their genera have a unique step-like ridge between the antennae.

Argia species have smooth faces and big postocular spots, yet are genetically closer to the ridge-faced groups. Other features, such as a pair of pillow-shaped knobs (called tori) above the male claspers, are unique to them. The genus is very distinct, therefore (and thus placed in its own subfamily), and could even have originated before coenagrionids invaded (and came to dominate) still waters.

As the same seems true for Protoneurinae (next profile), that implies that their habitat preference may actually be ancestral and shared with the pond-damsels' sister-group, the stream-damsels (p. 166)! Both groups might rather be seen as the New World equivalent of those, therefore. Indeed, most platycnemidids have long leg spines too, while some were once thought to be closely related to *Argia* (pp. 174 and 177). North America's Powdered Dancer (*A. moesta*) is stunningly like Africa's *Mesocnemis* (p. 172), for example.

ABOVE | The subspecies *violacea* of the Variable Dancer (*Argia fumipennis*); two subspecies in southern North America have entirely dark wings.

NEOTROPICAL THREADTAILS

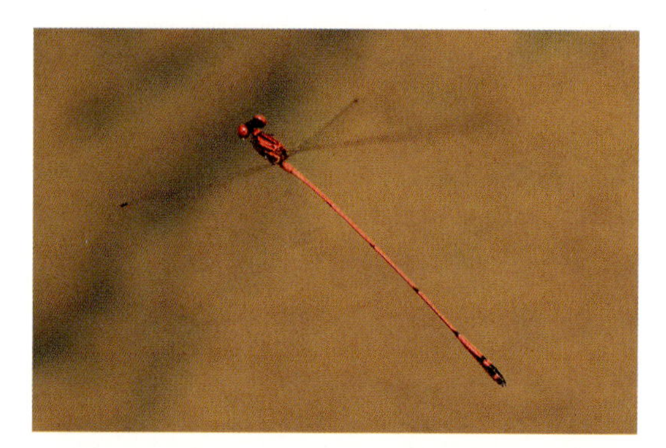

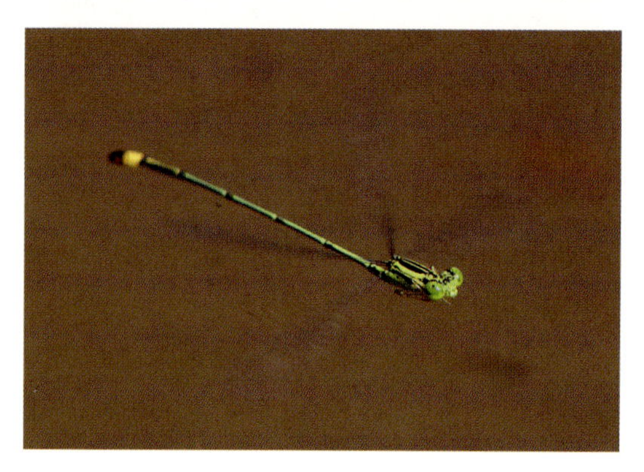

ABOVE | Males of the Crimson Threadtail (*Protoneura amatoria*, top) and Lemon-striped Threadtail (*Neoneura confudens*, lower) hovering over a stream in Panama.

While the smaller damsels on tropical streams associated with aquatic vegetation are of rather average build (p. 150) and those linked to stronger currents more robust (p. 152), threadtails represent a third archetype: with extended and extremely attenuated bodies, they are tied particularly to wooded, shaded waters.

Protoneurines are found from montane seeps and swamp forest pools to sluggish lowland rivers, typically bordered by trees. Brazil's tiny Eryngo Pincertip (*Roppaneura beckeri*) breeds in the water-holding leaf axils of a thistle-like weed, but most nymphs live among organic sediments such as leaf-litter. Males hang from leaf tips nearby but also hover above the breeding habitat for long periods.

The 67 quite similar species in *Amazoneura*, *Drepanoneura*, *Epipleoneura*, *Epipotoneura*, *Forcepsioneura*, *Peristicta*, *Phasmoneura*, and *Roppaneura* are very fine and generally dark and dull with pruinosity at the extreme abdomen end inspiring the name graytips; some with long forcipate male claspers are called pincertips. Brazil's leafhangers (*Idioneura*) and some of the delicate threadtails (*Psaironeura*, *Lamproneura*) from Amazonia to Central America have bright red thoraces and abdomen tips, common in other ridge-faced coenagrionids too (p. 158).

DIVERSITY
130 species of mostly running and often forested waters in the American tropics

TAXONOMY
Genera *Amazoneura* (3 species), *Drepanoneura* (8), *Epipleoneura* (30), *Epipotoneura* (2), *Forcepsioneura* (13), *Idioneura* (3), *Lamproneura* (1), *Neoneura* (29), *Peristicta* (8), *Phasmoneura* (2), *Protoneura* (22), *Psaironeura* (6), and *Roppaneura* (1) seem closely related, but *Junix* (1) and *Proneura* (1) may be nearer Ischnurinae (p. 144)

The slender threadtails (*Protoneura*) have elaborate red, orange, yellow, or blue markings and often brilliant red eyes. Some species seem to form leks, males hovering together among trees near their habitat where females can find them. Hovering over sunnier streams and rivers, the robust threadtails (*Neoneura*) are even brighter and rather short-bodied, so might better be called hoverdamsels or floaters.

As they hover, the threadtails' long abdomens probably provide balance. This flight style, moreover, seems linked to narrowed wings, in which the vein nearest the trailing edge (called the anal vein) is often reduced. In the past, this feature was used to define Protoneuridae as a family. Many tropical damsels, however, have this peculiar character not due to being related but to sharing similar habits.

Today, the remaining one-fifth of what was once one of the largest odonate families has been demoted to a subfamily of Coenagrionidae. Meanwhile, *Allocnemis* (p. 172) and Disparoneurinae (p. 170) were moved to Platycnemididae and *Lestoidea* to Lestoideidae (p. 207), and Isostictidae (p. 178) and Platystictidae (p. 233) became separate families.

Might some genera still be misplaced? Just a few specimens of *Junix elumbis* and *Proneura prolongata* are known from northwest Amazonia's blackwaters. Despite prolonged bodies and shortened anal veins, they recall equally elongate "leptobasines" such as *Mesoleptobasis* (p. 145). If that relationship is proven, the vein would indeed not even identify Protoneurinae within Coenagrionidae!

BELOW | A pair of the Crimson Threadtails (*Protoneura amatoria*) about to lay eggs.

HELICOPTERS

Helicopters seem to row through the air, dragging the longest bodies of all Odonata along on the longest wings. Beating slowly and often independently, the wings' whirling motion is emphasized by dense fields of white, yellow, red, or black cells in their broadly rounded tips (see photo p. 33). Replacing the stigmas, these markings are called pseudostigmas, and for almost a century their owners were treated as a separate family, Pseudostigmatidae.

Genetics proved, however, that these 27 giants evolved from among the over 400 ridge-faced coenagrionid species, treated next. How could they become so large while breeding in the smallest waters, such as tree-holes and bamboo stumps filled with rainwater? The less space a nymph has, the more advantageous it is to be big—to compete with other nymphs and also to cannibalize its kin!

LEFT | Up to 6 ¼ in (16 cm) long, *Mecistogaster* species such as this male Long-tailed Helicopter (*M. linearis*) from Panama are the longest odonates.

TOP RIGHT | Measuring up to 7 in (18 cm), the widest odonate wingspans are found among the *Megaloprepus* species, such as this male Blue-winged Helicopter (*M. caerulatus*) in Panama.

DIVERSITY
27 species breeding in tree-holes, bromeliads, and similar containers in the American tropics

TAXONOMY
Genera *Anomisma* (1 species), *Mecistogaster* (9), *Megaloprepus* (4), *Microstigma* (3), *Platystigma* (8), and *Pseudostigma* (2) placed formerly in Pseudostigmatidae seem closely related, but affinities with other ridge-faced genera (pp. 158–164) are unclear, so subfamilies such as Pseudostigmatinae and Teinobasinae cannot be delimited yet

ABOVE | Blue-winged Helicopter (*Megaloprepus caerulatus*) flying in the rainforest of Costa Rica.

For example, while the related Dow's Dryad (*Pericnemis dowi*) on Borneo produces several smaller individuals in large tree-holes, just a single large damselfly emerges from smaller ones.

Although ischnurines made a similar shift twice (pp. 140 and 143), all other times that coenagrionids moved into these water-pockets (at least eight incidences) ridge-faced groups were involved, probably because these were already adapted to shallow water with much detritus but little oxygen in their swampy forest habitats (compare p. 41): their nymphs often have expanded gills, which can be held against the water surface.

While some remain diminutive (p. 154), most pocket-breeders are much larger than their relatives nearby (see also pp. 158 and 198). Many favor particular plants: pandans are preferred by the related Philippine *Pandanocnemis* and New Guinea's *Papuagrion*, and bromeliads by the bromeliad-guards (*Leptagrion*) in Atlantic South America, the bromeliad-damsels (*Diceratobasis*) on Hispaniola and Jamaica, and Amazonia's *Bromeliagrion* (p. 161).

That helicopters evolved to even greater sizes may be linked to their conspicuous wing markings (compare p. 180) and feeding specialization: using their calm flight they pluck spiders and their prey from their webs. Waiting for rain to replenish their habitats, adults may live for over eight months. Females climb inside these confined spaces, awkwardly bending their long abdomens to place the eggs above the water. These hatch when, with the first downpours, temperatures drop. The nymphs may have bright spots on their upperside, perhaps to warn prospective mothers that the spot is already occupied.

COENAGRIONIDAE—REMAINING RIDGE-FACED GENERA OF TROPICAL FOREST

FINELINERS, SPINYNECKS, AND POSSIBLE KIN

Confined to Kenya and Tanzania's coastal forests, the East Coast Giant (*Coryphagrion grandis*) seems like a distant cousin of America's helicopters (previous profile). Indeed, Africa's largest damsel uses the smallest habitats, even including coconut husks and the shells of giant snails. Adults exclusively raid spiderwebs, moreover! This up to 4⅓ in (11 cm) beast's nearest relative is barely 1½ in (4 cm) long, however, and surprisingly nearby.

Found at high-elevation streams, the habitat of southern Malawi's Mulanje Damsel (*Oreocnemis phoenix*) could not differ more. It oviposits above water too,

TOP LEFT | The red colors and slender build of this male Indian Ocean Fineliner (*Teinobasis alluaudi*) in Madagascar are typical of many species in this group.

LOWER LEFT | Ranked as Critically Endangered on the IUCN Red List, the Mulanje Damsel (*Oreocnemis phoenix*) symbolizes the clear waters that flow down from its mountain home.

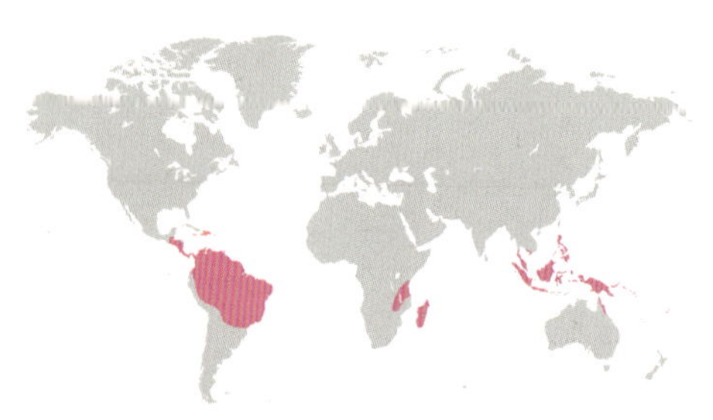

DIVERSITY
257 species found of tropical forests in the Americas, Asia, and Pacific, breeding in streams, pools, and small containers such as tree-holes

TAXONOMY
Genera *Coryphagrion* (1 species) and *Oreocnemis* (1) appear close, as do the "teinobasine" *Amphicnemis* (22), *Luzonobasis* (1), *Melanesobasis* (7), *Pandanobasis* (4), *Papuagrion* (29), *Pericnemis* (13), *Plagulibasis* (2), *Sangabasis* (12), and *Teinobasis* (88); affinities of *Aceratobasis* (4), *Bromeliagrion* (3), *Dactylobasis* (1), *Diceratobasis* (2), *Hylaeonympha* (1), *Inpabasis* (5), *Leptagrion* (20), *Metaleptobasis* (32), and *Tepuibasis* (9, including *Austrotepuibasis* and *Pseudotepuibasis*) are less clear

though, even at dry sites, apparently in anticipation of rain. Both species' habitats are also so marginal that they almost lack competitors. Circled by rockfaces over 3,200 ft (1,000 m) high, Mulanje is the highest massif within 900 miles (1,500 km). The East Coast Giant, meanwhile, survives only in a narrow strip reached by both northern and southern monsoons.

ABOVE | A pair of bromeliad-guards (*Leptagrion dispar*) in Brazil balancing on a leaftip of the tank bromeliad in which they will lay their eggs.

Probably, related species once inhabited the rivers and ponds nearby, but as constant change favored more competitive odonates (p. 20) only these two specialists were peripheral and also sheltered enough to survive. Today, little coastal forest remains, while mining looms over Mulanje, so both species are threatened with extinction.

Overall, the ridge-faces are much more localized than other coenagrionids. With exceptions, they are not only scarce in Africa, but also in temperate regions and exposed waters (but see p. 162), as well as faster-flowing ones (p. 154).

Most inhabit small, shaded waters, typically (tropical) forest pools and runnels. Interestingly, preferences for either running or standing water seem less clear-cut than generally in Odonata (p. 12), perhaps because oxygen levels (lower in warmer and more stagnant water) are less important overall. This, indeed, may also be the reason for their relative success in pocket-sized habitats (p. 157).

Some 130 described species from Malaysia to the Pacific, but concentrated on the islands of Indonesia, the Philippines, and New Guinea, form the heart of the possible subfamily Teinobasinae. Almost 90 are classified as fineliners (*Teinobasis*), but as the pandan-breeding *Papuagrion* (p. 157), as well as New Guinea's *Plagulibasis* and Fiji's *Melanesobasis* fall within their variation, that arrangement must be reviewed.

All are very slender, found in dark corners perched on leaf tips with the abdomen sloping down. Males are mostly black marked with various colors, a bright red thorax and abdomen tip being most typical. The Indian Ocean Fineliner (*T. alluaudi*) apparently crossed the sea to colonize the Seychelles, Madagascar, and even East Africa (compare p. 79), so perhaps the ancestors of *Coryphagrion* and *Oreocnemis* arrived thus too!

The Asian swampdamsels are even more fragile, males having distinctive green-metallic bodies and pale exaggerated claspers, while females are often startlingly all-red on the thorax and legs. Of the 52 very localized species, most that are limited to Sundaland remain in *Amphicnemis*, while many found from forest streams to tree-holes in the

Philippines were recently transferred to *Luzonobasis*, *Pandanobasis*, *Pericnemis*, and *Sangabasis*.

Over 75 species with comparable habits and habitats are known from tropical America. With long orange-tipped abdomens and reddish thoraces, spinynecks (*Metaleptobasis*) are amazingly fineliner-like. They are also remarkably like *Mesoleptobasis* and other "leptobasines" found in similar habitat, although the two spines in their "necks" project from the thorax's front rather than the prothorax's back (p. 145).

Most other species are known from few specimens or remote sites. Pendants (*Aceratobasis*) are limited to Brazil's Atlantic coast, for example, the sole *Dactylobasis* to the isolated Chocó on Colombia's Pacific side, and various *Tepuibasis* to bogs atop Venezuela's famed table mountains (p. 228). Scattered around Amazonia, *Hylaeonympha* and *Inpabasis* seem similarly rare, while *Bromeliagrion*, *Diceratobasis*, and *Leptagrion* are all large (but not gigantic!) bromeliad specialists (p. 157).

ABOVE LEFT | This green-metallic Wispy Swampdamsel (*Amphicnemis gracilis*) from Singapore is a male, as shown by the claspers and secondary genitalia at the tip and base of the abdomen (p. 17).

ABOVE | This very different-looking young *A. gracilis* is a female, having an ovipositor at the abdomen tip (p. 18).

RED DAMSELS AND POSSIBLE KIN

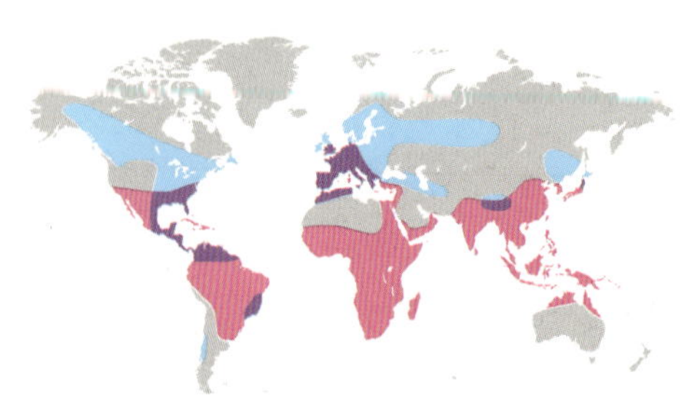

DIVERSITY

147 species of mostly sunny standing and slow-moving waters rich in vegetable matter worldwide

TAXONOMY

Genera *Aeolagrion* (4 species), *Angelagrion* (2), *Minagrion* (6), *Phoenicagrion* (7), *Schistolobos* (1), and *Telebasis* (62) seem close to each other, while *Chromagrion* (1), *Huosoma* (2), and *Pyrrhosoma* (2) certainly are; exact affinities of *Antiagrion* (4), *Ceriagrion* (50), and *Nehalennia* (6) unclear

Among the many blue-and-black damsels at sunny swamps and ponds in the warmer parts of the Americas, firetails (*Telebasis*) stand out with their wholly red abdomens. Coraltails and citrils (*Ceriagrion*) from Africa to Asia and north Australia similarly catch the eye. These quite closely related genera are clearly ecological counterparts, being among the few ridge-faced coenagrionids to frequent open stagnant waters.

Both favor the most developed aquatic vegetations, with mats of floating plants like duckweed, watermoss, and water lettuce being particular favorites. Some *Ceriagrion* like mossy bogs or forest pools full of detritus, as do Amazonia's cochadamsels (*Aeolagrion*) and flamechests (*Phoenicagrion*, *Schistolobos*) and Brazil's bluebands (*Angelagrion*) and bluetips (*Minagrion*), which are considered close to *Telebasis*.

Such microhabitats with much organic material, being warm but also quite oxygen-poor, may have provided a specific niche for these genera (**pink on map**) to diversify alongside the many other tropical coenagrionids. While both the New and Old World genera include largely blue, orange, or yellow males (some are even white or black), red may predominate as that color should stand out most in habitats that while quite open are also very lush (compare pp. 58 and 148).

Two *Ceriagrion*, known as small red damsels, are found in western Eurasia and north Africa. They and a montane species from southeast China have dark bronze thoraces and some other unusual features, so are sometimes separated as *Palaeobasis*. Aside from them, only 15 ridge-faced species in five genera inhabit temperate regions (**blue on map**); living inside dense vegetation such as peatmoss on the surface of bogs and seeps, the nymphs are buffered against cold in winter and warm up rapidly in spring.

Two large red damsels (*Pyrrhosoma*) also occur in western Eurasia and north Africa, with two equally big and red *Huosoma* species found in the Himalayas from Bhutan to southwest China. Inhabiting a broad range of standing and slow-flowing waters with dense submerged vegetation or leaf deposits, they are often the first damsels to emerge in spring.

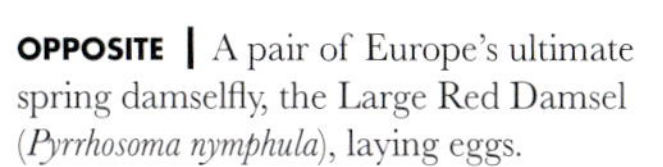

OPPOSITE | A pair of Europe's ultimate spring damselfly, the Large Red Damsel (*Pyrrhosoma nymphula*), laying eggs.

TOP AND LOWER RIGHT | These males of southern Asia's Orange-tailed Coraltail (*Ceriagrion auranticum*, top) and Brazil's Coral Firetail (*Telebasis corallina*, lower) are typical of their respective genera, which replace each other in the Old and New World.

BELOW | A male of the Orange-tailed Bluetip (*Minagrion waltheri*) from Brazil.

BOTTOM | A pair of Aurora Damsels (*Chromagrion conditum*), characteristic of the North American spring.

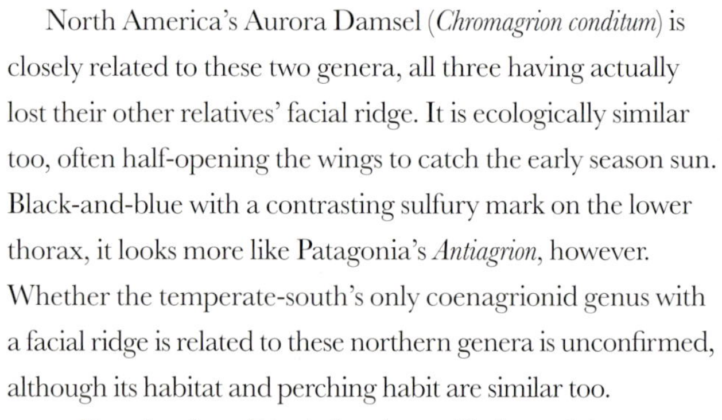

North America's Aurora Damsel (*Chromagrion conditum*) is closely related to these two genera, all three having actually lost their other relatives' facial ridge. It is ecologically similar too, often half-opening the wings to catch the early season sun. Black-and-blue with a contrasting sulfury mark on the lower thorax, it looks more like Patagonia's *Antiagrion*, however. Whether the temperate-south's only coenagrionid genus with a facial ridge is related to these northern genera is unconfirmed, although its habitat and perching habit are similar too.

Ranging from Rio de Janeiro to Alaska and then to Siberia and the Alps, the bog sprites (*Nehalennia*) do extend across the equator toward the Arctic: the Tropical Sprite (*N. minuta*) is largely blue marked with black, but northward the species' dark markings become increasingly expansive and strongly green-metallic, probably for thermoregulation (compare p. 76). Throughout, the teeny adults hide among grasses and sedges (see photo p. 132), the nymphs within dense plant matter.

COENAGRIONIDAE — *LEPTOCNEMIS*

SEYCHELLES STREAM-DAMSELS

The archipelago of the Seychelles is literally stranded in the middle of the Indian Ocean. When, between 150 and 70 mya, the supercontinent Gondwana broke up and the ocean widened (p. 111), some fragments stayed behind. Today, the granitic islands left above water lie over 600 miles (1,000 km) from Madagascar, 800 miles (1,300 km) from Africa, and 1,600 miles (2,700 km) from India. Two damselfly genera, each with just one species, occur nowhere else.

Of these, the Seychelles Islander (*Allolestes maclachlanii*) is the most disparate member of Podolestinae, a subfamily found also in Madagascar, central Africa, and Southeast Asia (p. 198). The Seychelles Stream-damsel (*Leptocnemis cyanops*) has no obvious relatives at all. Long placed in the Platycnemididae, treated next, its genetics and the male's spiny penis suggest it may represent the most isolated lineage of Coenagrionidae, having separated from them before all other 1,350 species arose, and thus before those came to dominate standing waters. A separate subfamily may have to be erected for it, therefore!

Seychelles' two specialties both rely on the few forested streams that remain on the islands Mahé, Praslin, and Silhouette. Their remarkably similar markings could be a coincidence, but might also provide some mutual (protective?) benefit.

DIVERSITY
1 species of forested streams on the three largest granitic islands of the Seychelles

TAXONOMY
Genus *Leptocnemis* (1 species)

ABOVE | The isolated Seychelles Stream-damsel (*Leptocnemis cyanops*) may be the most distinctive coenagrionid, probably deserving a subfamily of its own.

FAMILY: PLATYCNEMIDIDAE

STREAM-DAMSELS

As their name suggests, stream-damsels take the place of their nearest relatives, the pond-damsels (p. 132), in many of the Old World's (forest) stream and river niches. Nymphs live mostly among rootlets under banks, in deposits of detritus, or beneath rocks, rather than among water plants (compare pp. 152–5).

While not readily diagnosed as a whole, the combination of habitat with the adults' often bristly legs, hammerlike heads, and characteristic coloration patterns usually suffices to identify members of the family. Freshly emerged adults in some genera can be as white as ghosts, becoming marked

LEFT | This male Blue Featherleg (*Platycnemis pennipes*) in France hopes to mate with a female Common Bluet (*Enallagma cyathigerum*), a pond-damsel (p. 134).

TOP RIGHT | Males of many platycnemidids have expanded or colored legs and some, such as this Blood-red Featherleg (*Proplatycnemis sanguinipes*) from Madagascar, have both.

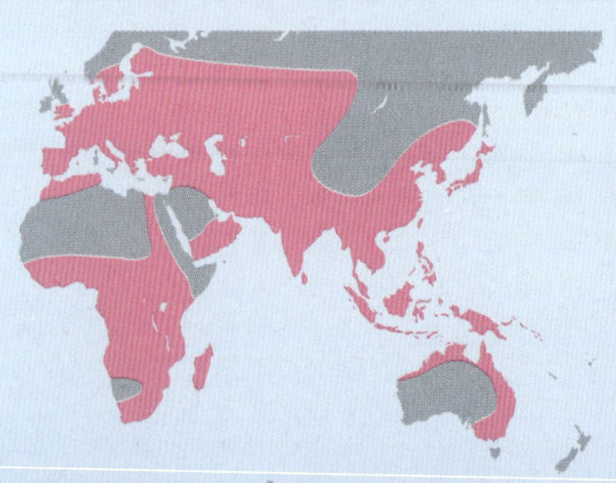

DIVERSITY
485 species found at most types of running (and some standing) waters from Africa to Australia

ADULT HABIT
Varied but mostly smaller damselflies recalling pond-damsels; in the warmer parts of the Old World, many damsels at flowing waters that rest with closed wings and have a simple square mesh of veins belong to this family, particularly those with a broad head and rather long, dense leg spines

TAXONOMY
6 well-defined subfamilies treated separately below

only gradually. Such pale females are often blushed orange or red, unlike their mature males and, indeed, most other damselflies.

While only one-twelfth of species has the expanded tibiae for which *Platycnemis* ("broad-shin" in Greek) was named (p. 168), genera such as *Allocnemis* (p. 172) and *Igneocnemis* (p. 175) must use their colorful legs similarly. Most of the many similar genus names refer to an affinity with the group, however: *Cyanocnemis* and *Macrocnemis* do not have big and blue tibiae, for example, but are big and blue themselves (p. 174).

RIGHT | A small nymph of the Blue Featherleg (*Platycnemis pennipes*) raising its leaflike gills.

PLATYCNEMIDIDAE—PLATYCNEMIDINAE

FEATHERLEGS

As they jostle for the best spots at the waterside, featherlegs wave their brightly colored and often very wide tibiae at each other threateningly. In mature *Platycnemis* males these quill-like legs are white, often with thin black shafts. The species are characteristic of slow-moving waters, especially open streams, rivers, and canals, but also sufficiently oxygenated ponds and lakes.

The western Eurasian species have abdomens in pastel shades of orange or blue, or even entirely white or pruinose. The East Asian ones are darker, more like the *Pseudocopera* species nearby, which inhabit smaller, shadier, and also more tropical waters; their narrower white legs are distinctively black ringed at the joints and appear even longer.

Unlike the tribe Platycnemidini, described above (which also includes Vietnam's poorly known *Matticnemis doi*), males of the strictly tropical Coperini can also have yellow, orange, red, blue, and even black legs. The nymphs have a unique fringe of long filaments on their three leaflike gills, moreover, while Platycnemidini at most have one filament at each

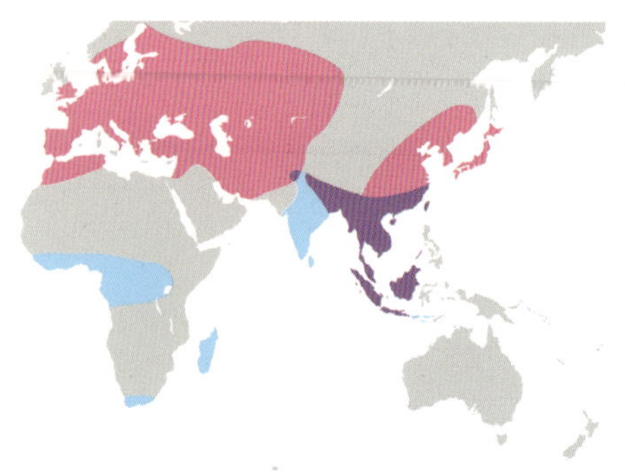

DIVERSITY
39 species at a broad range of mostly calmer waters from Europe and Africa to Southeast Asia

TAXONOMY
Tribe Platycnemidini (**pink on map**) with genera *Matticnemis* (1 species), *Platycnemis* (12), and *Pseudocopera* (4); and Coperini (**blue**) with *Copera* (9), *Proplatycnemis* (12), and *Spesbona* (1)

ABOVE | The quill-like legs of the Blue-faced Featherleg (*Proplatycnemis pseudalatipes*) male from Madagascar are especially wide.

TOP RIGHT | The Yellow Featherleg (*Copera marginipes*) is common across southern Asia.

gill's tip. Although some species prefer open streams and rivers, most inhabit forested and often muddy springs, drains, and pools.

Until 2001, only the genus *Copera* was known from Asia and Africa, while *Proplatycnemis* was limited to Madagascar and the Comoros. That year, however, *P. pembipes* was found at the only forest stream left on Pemba, off the Tanzanian coast. Genetics subsequently not only revealed that *Proplatycnemis* had been there for a long time, rather than blowing across recently, but that its nearest relative survived at Africa's extreme southern tip.

Not recorded since 1920, *Spesbona angusta* was rediscovered near Cape Town only in 2003. Its legs, while long, are narrow and entirely black. The nymph's gills too are fringed, each filament even being bizarrely branched! Previously placed incorrectly in *Metacnemis* (p. 173), its new genus name was inspired by the Western Cape's provincial motto "Spes Bona," Latin for "Good Hope."

Similarly to two other relict African damselflies (p. 158), Spesbona and the Pemba Featherleg only inhabit pools filled by seasonal streams, largely free of competing damselflies. This and their distribution on the continent's periphery suggest they may be the last survivors of a once widespread radiation of featherlegs.

ABOVE | Hugging a bright surface such as white sand or a dead stem in the sun, South Africa's small Spesbona (*Spesbona angusta*) male can turn from black to purple to blue in minutes.

PLATYCNEMIDIDAE—DISPARONEURINAE

PALEOTROPICAL THREADTAILS

Forming the main radiation of small damsels in the Old World tropics' running waters, besides *Pseudagrion* (p. 150), disparoneurines are the largest subfamily of Platycnemididae. Despite the males' distinctively shaped claspers (see photo top right), these threadtails were long misplaced with those of the New World in Protoneuridae (p. 155)

As most species have dark and slender bodies, with accents of color in males, they resemble *Protoneura*, particularly. Their bewildering variation commonly includes red to yellow notes, while patches of bright pruinosity are frequent (especially in Africa) and blue marks (mostly elsewhere) too. Colors can also differ between body parts, while adults are often largely pale at emergence, and all bright shades can (as in other groups) initially be pale blue.

The species' habitats are equally varied, from grassland rivers to boggy or rocky streams and muddy seeps in deep shade. Most favor slower-

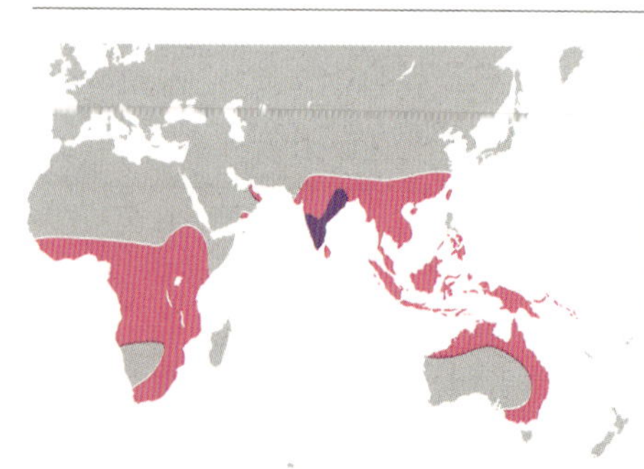

DIVERSITY

179 species at a broad range of running waters (but often forested and slower) from Africa to Australia

TAXONOMY

Tribe Disparoneurini (**pink and purple on map**) with genera *Arabineura* (1 species), *Disparoneura* (2), *Elattoneura* (45), *Nososticta* (89), and *Prodasineura* (38); and Caconeurini (**purple only**) with *Caconeura* (5), *Esme* (3), *Melanoneura* (2), and *Phylloneura* (2)

LEFT | A male Black-banded Threadtail (*Disparoneura quadrimaculata*) from India.

RIGHT | The short male claspers of the Common Threadtail (*Elattoneura glauca*) in South Africa are characteristic of the subfamily, with an often triangular tooth pointing down from the upper ones and (most distinctively) a narrow up-curled tip to the lowers.

BELOW | A male Coorg Bambootail (*Caconeura ramburi*) from India.

flowing and shadier waters, even in open landscapes. Despite their color, males are barely noticeable as they hover low over the water in dark places, or perch horizontally just above it.

At least 30 species placed in *Elattoneura* occur in sub-Saharan Africa, while 89 were already described in *Nososticta*, from Australia to the Andamans. Nearly 60 species, placed mostly in *Prodasineura*, range from Sri Lanka to Taiwan, the Philippines, and Greater Sundas; those retained in *Elattoneura* await taxonomic reassignment.

Despite their diversity in Oceania and Southeast Asia, the most distinctive species live on the Indian subcontinent, to which both *Disparoneura* are restricted. With an orange body and dark-spotted wings, the fairly robust Black-banded Threadtail (*D. quadrimaculata*) sits conspicuously on rocks in open rivers. The dull-pruinose Hajar Threadtail (*Arabineura khalidi*) may be related; it is limited largely to permanent springs in the desert mountains of Oman.

Bambootails are confined to the Western and Eastern Ghats. Hanging off leaf tips by (mostly shaded) springs and streamlets, their black bodies marked with blue, they appear to replace Southeast Asia's *Coeliccia* (p. 176) and Africa's *Allocnemis* (p. 172). While the 12 species seem distinct enough to have their own tribe, Caconeurini, their four genera differ only in size and thus venation.

With radiations rooted in Africa, Australia, and the Indian subcontinent, the subfamily's origins appear to lie in the supercontinent Gondwana (compare pp. 57 and 199). The three subfamilies discussed next follow a similar pattern, so perhaps the family as a whole originated there.

PLATYCNEMIDIDAE—ALLOCNEMIDINAE

YELLOWWINGS, RIVERJACKS, AND KIN

Slender and dark, yellowwings (*Allocnemis*) are hardly visible at their forest springs and streams. As they hover on amber wings in the deep shade, drifting spots of color—like notes highlighted with fluorescent marker—often draw our attention first to the males. The Rainbow Yellowwing (*A. nigripes*) is orange on the head and mint green on the thorax, for example, with a sky-blue line running toward its glaring orange abdomen tip.

TOP LEFT | With their white pruinosity, male riverjacks (*Mesocnemis*) such as this Savanna Riverjack (*M. singularis*) are conspicuous at rivers throughout tropical Africa.

LEFT | The tiny Powderblue Damsel (*Arabicnemis caerulea*) is less conspicuous in Arabia's desert valleys than its bright color suggests.

OPPOSITE | All *Allocnemis* males have bright marks on their dark bodies, but only South Africa's Goldtail (*A. leucosticta*) has such distinctive white stigmas.

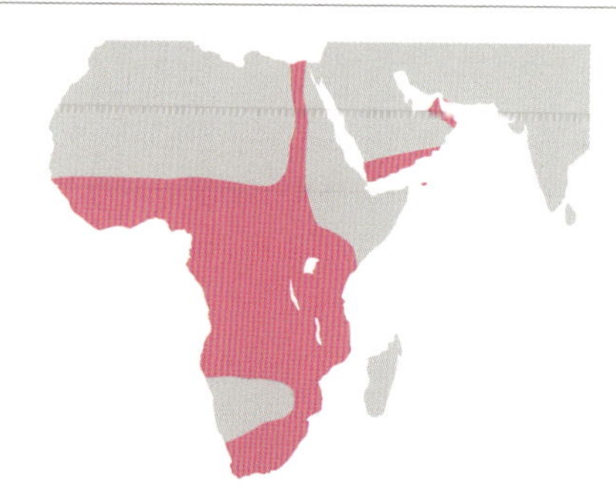

DIVERSITY

26 species at most running waters in tropical Africa and south Arabia

TAXONOMY

Genera *Allocnemis* (18 species), *Arabicnemis* (1), *Mesocnemis* (5), *Metacnemis* (1), and *Stenocnemis* (1)

While the above fit the threadtail archetype of stream-dwelling damsels (p. 154), riverjacks (*Mesocnemis*) recall the dancers in *Argia* (p. 152). Thickset and largely pale brown, males become almost completely white-pruinose except for their blue eyes. Rather than hovering cautiously and dangling off leaf tips, they sit horizontally on exposed snags and rocks on fast-flowing rivers or wave-battered lakeshores, skipping over the wild water from perch to perch.

Aside from these closely related but very different tropical African genera, just three are known, each with a single highly restricted species. The Blue Riverjack (*Metacnemis valida*) in South Africa's Eastern Cape is essentially a colorful *Mesocnemis*, with bright blue-and-black males and largely orange-brown females with creamy sides. In the rainforests around the Gulf of Guinea, the Tricklejack (*Stenocnemis pachystigma*) breeds in the most superficial streams, such as in the spray of waterfalls.

The Powderblue Damsel (*Arabicnemis caerulea*), finally, is the subfamily's only non-African member, found in the few wadis in southern Arabia with permanent streams. The uniformly pale blue male recalls a slight riverjack, but, being diminutive and rather inactive, can be hard to spot in its arid home.

BIRDLINGS AND RIPPLETIPS

With a surface area of France and Britain combined and peaks surpassing the height of the Alps, New Guinea is the world's highest and second-largest island. Stretching along the equator for almost 1,500 miles (2,500 km), its tropical biodiversity is unparalleled, the well-named birds-of-paradise being most famous. Presently, some 500 odonate species are known from here and almost 350 from nowhere else.

Between 1932 and 1987, Maurits Lieftinck described a staggering 280 of these for the first time. Seeking suitable names, he drew from the island's avian riches: the birdlings *Palaiargia alcedo*, *P. ceyx*, *P. halcyon*, *P. melidora*, and *P. tanysiptera* are named after kingfishers; *P. charmosyna*, *P. eclecta*, *P. eos*, and *P. micropsitta* for parrots and lorikeets; while *P. arses* denotes a monarch flycatcher and *P. myzomela* a honeyeater.

They belong to 83 species known currently from here and the adjacent Moluccas and Bismarck Archipelago. Possessing a bewildering range of male sexual characters (p. 13), these have been placed in no fewer than 12 genera. The *Palaiargia* species, for example, got their exotic

DIVERSITY
121 species found at most (forested) running waters on the Philippines and from the Moluccas to the Solomon Islands

TAXONOMY
Genera *Archboldargia* (3 species), *Arrhenocnemis* (3), *Cyanocnemis* (1), *Hylaeargia* (5), *Idiocnemis* (23), *Igneocnemis* (20), *Lieftinckia* (7), *Lochmaeocnemis* (1), *Macrocnemis* (1), *Palaiargia* (26), *Papuargia* (2), *Paramecocnemis* (5), *Rhyacocnemis* (4), *Risiocnemis* (18), *Salomocnemis* (1), and *Torrenticnemis* (1)

names for some truly startling combinations of red and blue markings, sometimes with orange and green as well.

Many species have very complex claspers too, those on the massive abdomen tip of *Archboldargia scissorhandsi* resembling the razor-tipped fingers of the movie character Edward Scissorhands. Male *Paramecocnemis* have odd hair tufts below both ends of their abdomens, which can be up to two and a half times longer than the wings.

OPPOSITE | Robust idiocnemidines such as this male Trauna Birdling (*Palaiargia traunae*) from New Guinea recall the distantly related genus *Argia*.

ABOVE | The magnificent male Red-hot Rippletip (*Risiocnemis praeusta*) from Samar and Leyte islands in the Philippines.

Aside from this spectacular radiation, eight *Lieftinckia* and *Salomocnemis* species are known from the Solomon Islands, while two genera of about 20 diversified on the Philippines's many islands. Male firelegs (*Igneocnemis*) vary from all red to wholly black, usually with the flaming red (but sometimes yellow, blue, or black) legs they are named for. Some rippletips (*Risiocnemis*) look similar, but many have dark legs or blue-spotted black bodies.

The latter's common name refers to its wingtip's finely serrated margin, which is diagnostic of the subfamily. These crenulations are absent in the genera with *argia* in their names; these were long associated with the coenagrionid genus *Argia* due to their stronger build (compare p. 152). Only three genera have been studied as nymphs, but their gills were found to be unique, with ruffed borders like leaves of kale.

PLATYCNEMIDIDAE—CALICNEMIINAE

SYLVANS AND OREADS

Of all odonate gems in Asia's rainforests, few invite a treasure hunt more than sylvans (*Coeliccia*). All are slender damsels marked subtly with (mostly) pale blue, bright yellow, and pruinose-white. Some are extensively yellow at first, changing to black and blue. Although up to nine species may occur nearby, many are very localized, so 30 were only described in the last 20 years.

As a group, calicnemiines are recognized by two elongate spots on the back of the head, not unlike coenagrionids' postocular spots (p. 153). Generally, the species prefer smaller and steeper running waters than nearby Disparoneurinae (p. 170); they are often found resting with half-open wings on leaves overhanging small streams, muddy springs, or wet rockfaces by waterfalls, predominantly in forest.

Most *Coeliccia* species occur in Indochina, extending to Taiwan and the Ryukyus, onto Sundaland, and along the Himalayas. Southeast Asia's *Indocnemis* only really differ by being bigger and beautifully marked violet-blue. Some distinctive species from Borneo and Java placed in *Coeliccia*, moreover, as well as all Philippine ones (from Mindoro, Mindanao, and Palawan), seem closer to those put in the Palawan genus *Asthenocnemis*. The large genus *Coeliccia* must clearly be revised, therefore.

Found predominantly on the mainland in cooler and more mountainous terrain, the oreads (*Calicnemia*) are more compact, marked extensively with flaming orange and brilliant scarlet, particularly the abdomen often being largely red. A few species are predominantly black, with gray-pruinose abdomens and markings instead.

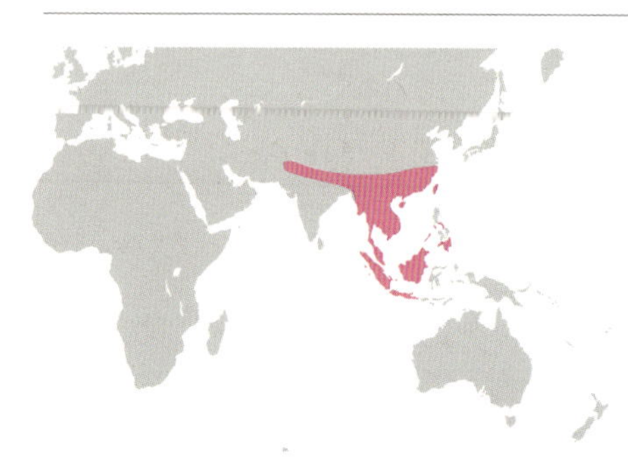

DIVERSITY
108 species of mostly smaller, forested running waters from the Himalayas to the Ryukyus, Philippines, and Java

TAXONOMY
Genera *Asthenocnemis* (2 species), *Calicnemia* (23), and *Indocnemis* (3) may fall within *Coeliccia* (80) as currently delimited

TOP | The pretty males of this group, such as this Fiery Oread (*Calicnemia erythromelas*) from Vietnam, are often found resting with half-open wings on leaves overhanging the tiniest trickles of water in Asia's forests.

LOWER | A male of Borneo's Upland Sylvan (*Coeliccia nemoricola*).

PLATYCNEMIDIDAE—ONYCHARGIINAE

SHORTTAILS AND WHISKERLEGS

The distinctive Black Shorttail (*Onychargia atrocyana*) is widespread in tropical Asia, inhabiting lushly vegetated stagnant waters, often in or near forest. While frequently abundant, adults are inconspicuous, often sitting high up in bushes and trees. Boldly yellow striped at emergence, males can become all black with a purple-glossed thorax. For a century and a half, this "Marsh Dancer" was considered its genus's only species and placed near the American dancers in Coenagrionidae (p. 152).

Only in 2015 was the White Shorttail (*O. priydak*) named from Cambodia and Thailand; its thorax and femora become thickly white-pruinose. Genetics bore out that *Onychargia* was nearer Platycnemididae than American dancers, moreover, but formed the most distinct subfamily therein. While its habitat and short abdomen suggested a pond-damsel, obscure details such as the spineless penis, a hidden tooth at the base of the male's tiny claspers, and the (vanishing) head markings indeed fitted the stream-damsels better.

The genetics also unearthed that Madagascar's whiskerlegs (*Paracnemis*) are the shorttails' only close relatives. While their markings are blue and do not obscure with age, their forest swamp habitat is quite similar, as are structural details. Both genera have long bristle-like spines on the legs, moreover. Those gave the whiskerlegs their name, but are also more typical of platycnemidids than coenagrionids.

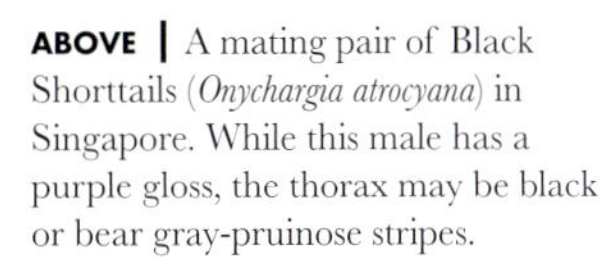

ABOVE | A mating pair of Black Shorttails (*Onychargia atrocyana*) in Singapore. While this male has a purple gloss, the thorax may be black or bear gray-pruinose stripes.

DIVERSITY
4 species of swampy standing waters, often in or near forest, in tropical Asia and Madagascar

TAXONOMY
Genera *Onychargia* (2 species) and *Paracnemis* (2)

FAMILY: ISOSTICTIDAE
WIRETAILS

LEFT | A female of Australia's Southern Pinfly (*Neosticta canescens*) showing its somewhat peculiar proportions.

RIGHT | A male of the vine-like *Selysioneura thalia* from the Moluccas.

Isostictids might be the least extravagant members of an insect order famed for its color and charisma. Males may have some metallic gloss, whitish pruinosity, or muted green, blue, or orange, but even they lack conspicuous marks of bright color. Many forest species have intricate, dark patterns, moreover, resembling camouflage. The adults are indeed easily overlooked. *Selysioneura* species hang vertically on the underside of leaves and twigs, for example, several meters above the water.

The adults appear disproportioned, and the rather stout nymphs are peculiar too. The three terminal gills, which can be leaflike or swollen, and sometimes very hairy, are uniquely constricted toward their end, like a joint or sausage link.

DIVERSITY
46 species of running waters and associated pools from Australia to New Caledonia and the Moluccas

ADULT HABIT
Small, slight damselflies with cryptic appearance and proportionately small, narrow head and thorax, weak small-spined legs, narrow (often pointy) wings, and long abdomen (comparatively thick in females)

TAXONOMY
Genera *Austrosticta* (3 species), *Cnemisticta* (2), *Eurysticta* (4), *Isosticta* (5), *Labidiosticta* (1), *Lithosticta* (1), *Neosticta* (3), *Oristicta* (2), *Rhadinosticta* (2), *Selysioneura* (16), *Tanymecosticta* (6), and *Titanosticta* (1)

Little is known about the species' habits and ecology, however. Over half inhabit rainforest streams on New Guinea and the nearby Moluccas, Solomon Islands, and New Britain (most placed in *Selysioneura*, but also *Cnemisticta*, *Tanymecosticta*, and *Titanosticta*), while *Isosticta* occurs at rocky forest streams and rivers on New Caledonia.

Sixteen species are confined to Australia. The wiretails (*Labidiosticta*, *Oristicta*, and *Rhadinosticta*) and pinflies (*Neosticta*) occur mainly at running waters in the humid east. The pondsitters (*Austrosticta*), pins (*Eurysticta*), and a single narrowwing (*Lithosticta*) are largely restricted to the far north's seasonal tropics. They also occupy ponds, such as those left in gorges by rain or along rivers.

Altogether, isostictids seem weak fliers and thus poor dispersers, but nonetheless persist in some tough environments. Perhaps their cryptic appearance allows the adults to survive unfavorable periods, while the eggs are often laid in vegetation above water, rather like the hardy Lestidae (p. 238). *Isosticta* species, by contrast, may even oviposit on boulders in rushing water and pouring rain.

Genetics indicate the family is closest to the great radiation of coenagrionid and platycnemidid damselflies, so it is usually included in the superfamily Coenagrionoidea (p. 132). Perhaps these modest Australasian species show us how small damsels looked and behaved before they conquered the world. Indeed, rather than having complex claspers like many of their nearest relatives, males retained the simpler forcipate structures seen in most of the families treated next.

FAMILY: CALOPTERYGIDAE

DEMOISELLES AND JEWELWINGS

While the often widespread and abundant damsels in the previous profiles are quite modest in appearance, the denizens of running waters treated next are among the largest and showiest. Most calopterygids have broad and finely reticulate wings that in males are often tinted, iridescent, or both, resulting in brilliant colors that flash like stroboscopes in flight.

Recalling small stick-insects, the spindly nymphs could hardly differ more, with their long antennae, masks, legs, and gills (see photo p. 9). They are well camouflaged among aquatic vegetation, submerged rootlets of bankside trees, or decomposing branches and leaves. This is also where the eggs are laid, and females may descend deep underwater to do so.

As such spots are patchy, it is worth defending them and having signals that attract mates and impress rivals (photo p. 33). This favors males with big wings and the large bodies and dense venation to support them. Perhaps such heavy, slow-beating wings do not require a stabilizing counterweight at their tips, as many calopterygids lack stigmas.

Once, Chlorocyphidae (p. 188), Euphaeidae (p. 204), Polythoridae (p. 218), and various "amphipterygid" (p. 209) and "megapod" groups (p. 194) were all united with this family within the superfamily Calopterygoidea. Study of nymphs and genomes revealed that showiness evolved many times in distantly related lineages of riverine damsels, however. Only Calopterygidae, the most widespread and species-rich group among these so-called "Caloptera" ("beautiful-wings" in Greek), is well-established beyond the tropics.

LEFT | This male Hainan Bluewing (*Matrona mazu*), found only on the island of Hainan, is an archetypical calopterygid with unstalked, colored, and densely veined wings lacking stigmas.

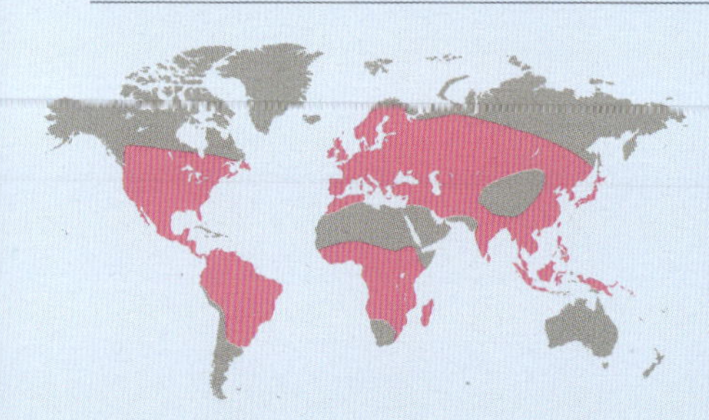

DIVERSITY
182 species of running waters worldwide, but absent from Australia and the Caribbean

ADULT HABIT
Mostly large damselflies, typically with metallic bodies and broad, reticulate, and often colored wings kept closed at rest; most damsels whose wings narrow gradually and are densely veined right up to their base (rather than appearing stalked) and also bear no (or reduced) stigmas belong to this family

TAXONOMY
2 subfamilies, Hetaerininae (p. 187) and Calopteryginae, the latter with 7 well-defined tribes treated separately in the following profiles

TYPICAL DEMOISELLES AND JEWELWINGS

While the subfamily Hetaerininae is strictly American (p. 187), the roots of the more varied Calopteryginae lie in tropical Asia, with four of its seven tribes only present there, along with most of the diversity of Calopterygini. Only the genus *Calopteryx* (**pink on map**) extends into northern Eurasia (where its species are commonly known as demoiselles) and North America (jewelwings). The metalwings (*Neurobasis*) similarly swept from India across the Indo-Pacific to New Guinea and the Solomon Islands (**dark pink**). Although they fell just short of Australia, this spread is exceptional for a family that is conspicuously absent from many islands.

Among southern Asia's most familiar odonates, Green Metalwing (*N. chinensis*) males have clear forewings, but green-metallic hind wings with sharply contrasting black tips. Females have strange white false stigmas near their brownish wings' tips, like other family members, but at the nodes as well. Other species' males have wholly violet- or blue-metallic wings, recalling Borneo's impressive Mountain Metalwing (*Matronoides cyaneipennis*). Most bluewings (*Matrona*) and the very close jetwings (*Atrocalopteryx*) occur in mainland China, extending across to Japan, Taiwan and Hainan (**blue on map**).

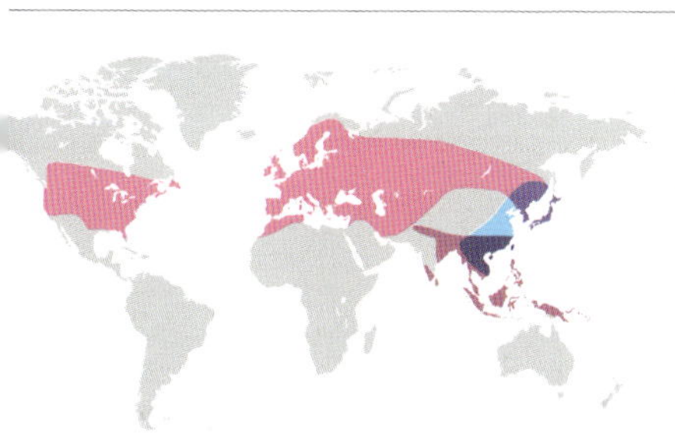

DIVERSITY
47 species of running waters from North America to the Solomon Islands

TAXONOMY
Genera *Atrocalopteryx* (8 species), *Calopteryx* (16), *Matrona* (9), *Matronoides* (1), and *Neurobasis* (13)

TOP | Two or more dazzling *Calopteryx* species, such as this male Beautiful Demoiselle (*C. virgo*) in Germany, are often common on the same streams and rivers in North America, Europe, and Japan.

LOWER | The Green Metalwing (*Neurobasis chinensis*) is among southern Asia's most familiar odonates.

COPPERWINGS, ECHOES, AND KIN

With deep amber wings and white-dusted bodies, male copperwings (*Mnais*) seem quite different from typical jewelwings (previous profile). The green-metallic females with densely veined see-through wings, however, have the basic calopterygid look, and occasional males appear exactly the same. Lacking their bigger brothers' flashy wings and reflective pruinosity, these so-called sneakers do not defend a territory, but try to intercept females when territorial males are distracted. That seems a poor strategy, but, by exerting themselves less, they live longer, ultimately attaining similar reproductive success!

Found from Japan to Indochina, all *Mnais* species can have both forms. Most prefer half-shaded streams, but where two meet one may move

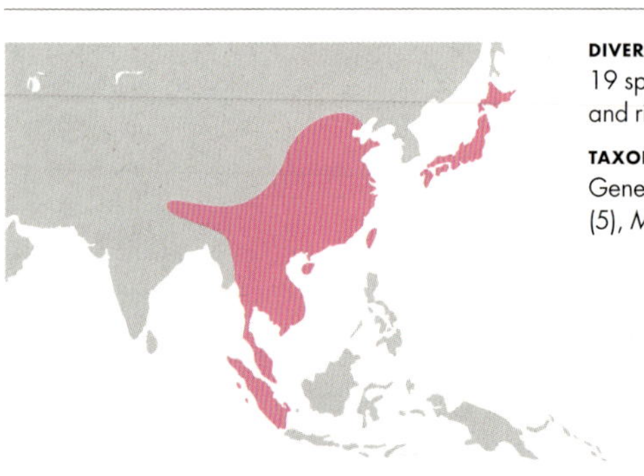

DIVERSITY
19 species mostly of forested streams and rivers in Southeast Asia

TAXONOMY
Genera *Archineura* (3 species), *Echo* (5), *Mnais* (9), and *Psolodesmus* (2)

ABOVE | A brightly colored male of the Japanese Copperwing (*Mnais costalis*) out in the sun.

RIGHT | With its massive thorax that dwarfs the head, southern China's 3½ in (9 cm) long Rosy Blossomwing (*Archineura incarnata*) recalls a bodybuilder.

into the sun, the other to the shade. And while the sun-dwelling species can just have amber-winged males, which get even larger, the shade-dweller only has clear-winged and smaller males. Apparently, it's deleteriously confusing for close relatives to look too much alike! Such "character displacement" may be a major evolutionary force in showy odonates (compare pp. 184 and 187).

All the tribe's species are characterized by bold patterns of pruinosity. Male *M. gregoryi* from southwestern China, for example, have the middle of their wings black, so the white-pruinose outer thirds stand out even more. Perching in deep shade, the dark-bodied and clear-winged *Echo modesta* males (one of five echoes from the eastern Himalayas to Sumatra) draw all attention to their incandescent white faces.

Males of the closely related Formosan Piedwing (*Psolodesmus mandarinus*) have dark wings traversed by a broad subapical white band in north Taiwan, but purple-shimmering dark-tipped wings in the south. On the nearby Ryukyu islands, *P. kuroiwae* has smaller dark tips but bigger, distinctly white-pruinose stigmas. The males of southern China's spectacular Rosy Blossomwing (*Archineura incarnata*) have crimson wing bases shot through with white-pruinose veinlets, and thus startlingly pink. The Cloudy Blossomwing (*A. hetaerinoides*) from Yunnan and bordering Laos and Vietnam is even bigger, and flushed white instead.

These are the largest calopterygids and must owe their muscular build to the males' high-powered contests, which take them soaring out of sight. The most superb *Archineura* may have definitively ascended to the heavens, however. Just one female of the Lost Blossomwing (*A. maxima*) is known, caught in north Vietnam in 1901. Females of the two other species have clear wings, but hers have semitranslucent cream bases, trimmed by dark bands. What might the male have been like?

CALOPTERYGIDAE—CALOPTERYGINAE—REMAINING ASIAN TRIBES

FLASHWINGS, BRONZEBACKS, AND SLENDERWINGS

Found from India to Borneo, flashwings (*Vestalis*) are among Asia's more modest calopterygids. Often remaining in the forest shade, males are green-metallic with violet-sparkling glasslike wings. Both the wings and bodies of Java's Nila Flashwing (*V. luctuosa*), Sumatra's *V. lugens*, and the Philippines' *V. melania* are wholly dark but sparkle brilliant blue, like many Calopterygini (p. 181). They and the Indian and Sri Lankan Black-tipped Flashwings (*V. apicalis* and *V. nigrescens*) occur at the periphery of the greatest diversity of both the genus and family, however, suggesting that this group can only attain a showy appearance where fewer other showoffs are present (compare p. 183).

At the other edge of their range, higher up in Indochina and southern China, the genus

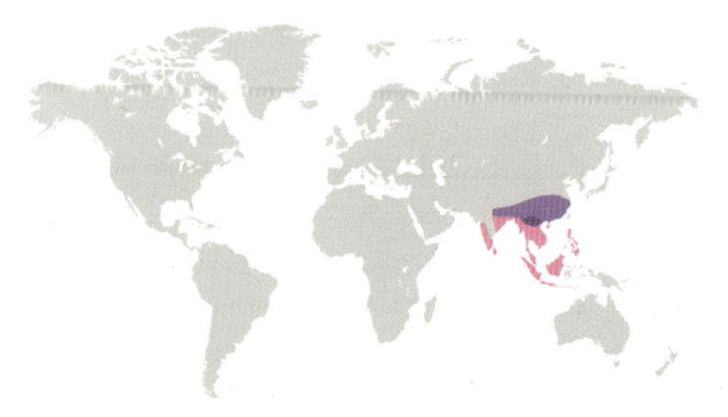

DIVERSITY
3 distinct tribes with 30 species from mostly forested streams and rivers in South and Southeast Asia

TAXONOMY
Tribe Caliphaeini (**purple on map**) with genus *Caliphaea* (6 species); Noguchiphaeini (**dark purple**) with *Noguchiphaea* (3); and Vestalini (**pink and purple**) with *Vestalaria* (5) and *Vestalis* (16)

ABOVE | Thai Slenderwing (*Noguchiphaea yoshikoae*) male.

OPPOSITE TOP | Common Flashwing (*Vestalis amethystina*) male from Singapore.

RIGHT | Thai Bronzeback (*Caliphaea thailandica*) male.

Vestalaria—only distinguished from *Vestalis* in 2006—is also rather distinctive. With the yellow underside of their thorax, pruinose tail-end in males, and often brown- or amber-tinted wings in both sexes, the species recall the co-occurring *Mnais* (p. 182). They appear in the fall, however, rather than early spring, and favor deep shade, especially bamboo thickets. Perhaps their similarity reflects shared adaptations to a comparatively cool environment and season, therefore.

The least conspicuous calopterygids, finally, are the bronzebacks (*Caliphaea*), which range from Nepal to southeast China, and Indochina's slenderwings (*Noguchiphaea*). Both look like very narrow-winged *Vestalaria* with at most a small dark dot at the *Noguchiphaea* forewing tip. As hinted by their genus names, they also recall such slender euphaeids as *Cryptophaea* and *Schmidtiphaea*, which are found at similar mountain forest streams (p. 206). Despite their superficial similarity, genetics suggest that Caliphaeini and Noguchiphaeini represent distinct (possibly ancient) tribes from Vestalini. Little is known about them, although *Noguchiphaea* oviposit in the bark of overhanging branches, rather than in softer organic substrates underwater like most calopterygids (p. 180).

SHINYWINGS, SPARKLEWINGS, AND KIN

Just as Calopteryginae probably expanded from Asia to Melanesia, and across Eurasia to North America, only once (p. 181), the Afrotropics and South America were colonized only on single occasions. The latter events, however, took place much longer ago; the two groups are more distinct therefore and classified as unique tribes today.

Merely a tenth of calopterygid species occur in the African tropics, with most in the continent's forested heart. The broadwings (*Sapho*) have (partly) blue-glossed black, almost paddle-shaped wings; they probably evolved from the rather modest sparklewings (*Umma*), named for their translucent wings' metallic glint. While the Emerald Demoiselle (*Phaon camerunesis*) is limited to rainforests, the Glistening Demoiselle (*P. iridipennis*) occurs everywhere else in tropical Africa, even tolerating seasonal flows. Waiting beneath dense vegetation, the dull and lanky adults are virtually invisible, but when flushed, patches of green iridescence on their forewings flash with every beat. The similar Madagascar Demoiselle (*P. rasoherinae*) is the only calopterygid to have reached the island (p. 230).

Calopterygids may once have extended from Eurasia into Africa and from there to South America. Fossils recalling Saphoini are known from Europe, while the shinywings (*Iridictyon*) of Venezuela's tepuis (p. 228) are not dissimilar. Ultimate proof, however, is yet to be found.

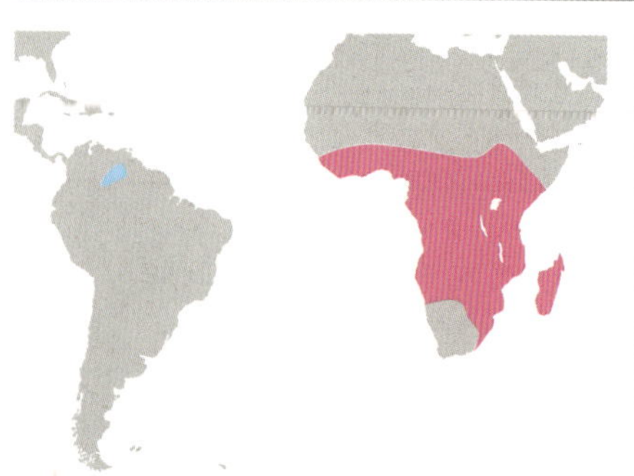

DIVERSITY
2 distinct tribes of mostly forested streams and rivers: 1 with 19 species confined to tropical Africa and Madagascar, the other with 2 in northern South America

TAXONOMY
Tribe Iridictyonini (**blue on map**) with genus *Iridictyon* (2 species); and Saphoini (**pink**) with *Phaon* (3), *Sapho* (6), and *Umma* (10)

TOP | Found mainly in Cameroon, the Purple Sparklewing (*Umma purpurea*) is among Africa's most beautiful species.

LOWER | Broadwings (*Sapho*), like this male Spring Broadwing (*S. bicolor*) from Gabon, are among the family's boldest and bulkiest members.

RUBYSPOTS AND METALLICS

While many calopterygid males perform displays to persuade females to mate and oviposit in their territory, others claim areas only as a rendezvous, as females arrive with their minds made up and mating and egg-laying occur elsewhere.

Hetaerinine males are indeed comparatively unshowy, with narrow wings, though over half the species have conspicuous red patches at their bases, often with dense reticulation of veins. Frequently, the wingtips bear small red spots too, while some species have extensively sooty wings. These so-called rubyspots (*Hetaerina*) inhabit flowing waters from Oregon and Quebec to Buenos Aires. Any given stream usually has just one or two species, but you will often find more at similar spots nearby.

The species from tropical South America lacking red in the wings have been dubbed metallics and placed in different genera, although genetic studies suggest all are just rubyspots that lost their ruby spots, perhaps due to character displacement (compare pp. 183–4), and so should possibly move to *Hetaerina* too. Of these, *Ormenophlebia*, with very elongate males (3–3⅛ in/7.5–8.0 cm versus 1½–2½ in/4.0–6.5 cm) and very long and clear wings, is restricted to cold streams in the Andes, the similar *Bryoplathanon globifer* to the podocarp and araucaria forests of Atlantic Brazil. The remaining species, with beautifully red- or purple-metallic (or wholly gray-pruinose) males, occur mainly in Amazonia and are placed in *Mnesarete*.

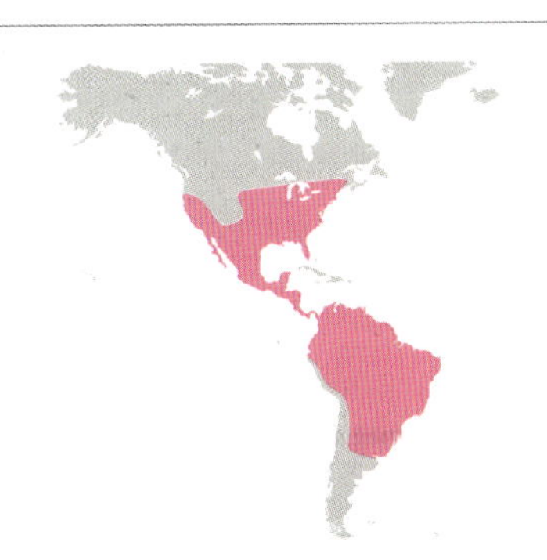

DIVERSITY
65 species of running waters in the Americas, although only 3 occur north of Mexico and none in the Caribbean

TAXONOMY
Genera *Bryoplathanon* (1 species), *Hetaerina* (37), *Mnesarete* (23), and *Ormenophlebia* (4) are under review

TOP | A male of the Racket-tipped Rubyspot (*Hetaerina occisa*) showing its diagnostic red markings.

LOWER | *Mnesarete* species such as this one from Peru can be hard to separate, differing mainly by the male's appendages.

FAMILY: CHLOROCYPHIDAE

JEWELS

With their males' gaudy attire, big and often bright noses, engaging dance routines, and sometimes even oversized feet, chlorocyphid adults might be seen as the clowns of the odonate world. They certainly form the most recognizable damselfly family, and (despite their size and resting posture) might be mistaken for dragonflies. The quite robust nymphs are equally unique, recalling earwigs with two long and spikelike appendages on the abdomen tip.

These nymphs live among roots, detritus, and silt, females mostly placing their eggs in dead leaves, sticks, and wood lying in the current nearby. Males converge at suitable spots, a focus that has led to an array of reproductive behaviors and ornamentations that make the calopterygids seem brash and unrefined (p. 180). Rivals are engaged frantically, typically by hovering opposite or beside each other, while often seesawing upward into the canopy. When the drab females approach, they are courted with aerial dances. Some species display while perched, vibrating their fanned wings like a peacock.

All these behaviors are highly ritualized; blue faces, white leg streaks, magenta wedges on the thorax, bronzy wings, ruby abdomens, and all sorts of other colorful signals, often on expanded surfaces, are flashed or wiggled rapidly. Such eccentric displays require erratic flight, probably made possible by the absence of the long, stabilizing abdomen found in other damselfly families.

LEFT | A male Red-nosed Gem (*Libellago sumatrana*) from Sumatra demonstrating the family's distinctive short body and swollen face.

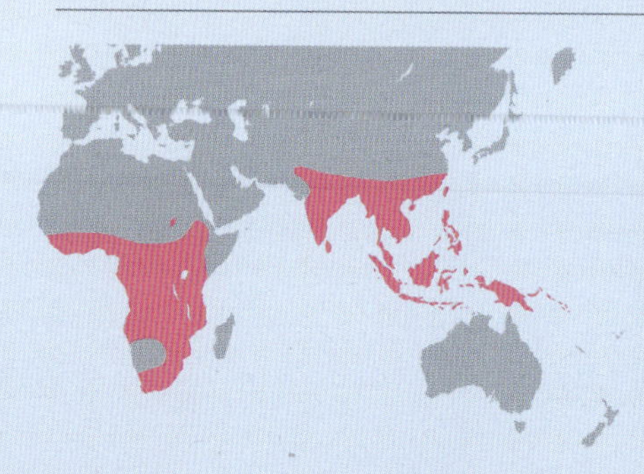

DIVERSITY
162 species of streams and (even large) rivers, as well as some seeps and rocky lakeshores, from the tropics of Africa to Melanesia

ADULT HABIT
Very small to medium-sized damselflies with swollen faces, bulging eyes, and short abdomens that rarely reach past the closed wings at rest; the wings are rather narrow yet densely veined with rather long stigmas and the male body (and often wings) typically brightly marked

TAXONOMY
Subfamily Chlorocyphinae restricted to Africa (see opposite); precise delimitation of the 2–5 subfamilies elsewhere (as well as some issues with the precedence of their names) are unresolved, so a definitive classification cannot be presented

JEWELS OF AFRICA

Despite the different ways the males show off (see opposite) genetics indicate that calopterygids and chlorocyphids are each other's closest surviving relatives. Both families are most diverse in tropical Asia, reached New Guinea but not Australia, and invaded the Afrotropics only once.

While the African males' wings play a minor role in display, invariably being clear, their faces, legs, and abdomens are often brightly colored and may even be broadened for optimal effect. Most *Platycypha* species, for example, have expanded tibiae recalling those of featherlegs (p. 168). While they occur mainly in eastern and southern Africa, often in open and elevated habitats, the typically broad-bodied *Chlorocypha* dominate the low-lying forests of the center and west. Five species from central Africa with slender legs and abdomens form the distinct genus *Stenocypha* (see photo p. 14).

Largely restricted to Cameroon and Gabon, *Africocypha* species may have the strangest color variation in Odonata. The abdomen of the Kaleidoscope Jewel (*A. lacuselephantum*) turns from pale blue to orange to bright red, but as this starts at the tip, then the base, and ends in the middle, the male will at some point have all three colors at once, on contrasting segments. While females are normally especially demure in this family, they are many-colored in this genus too.

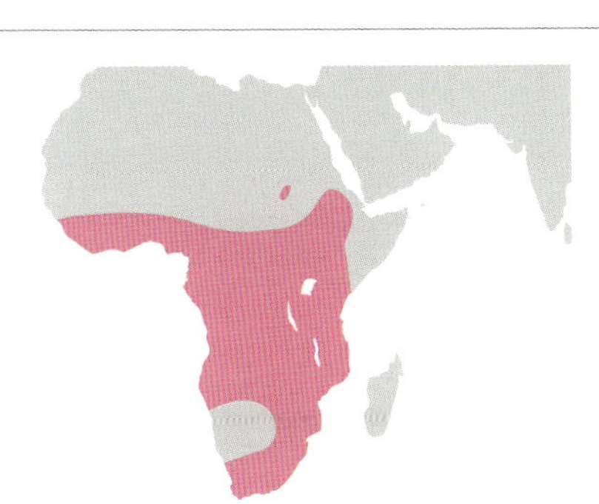

DIVERSITY
50 species of streams, rivers, and lakes in tropical Africa

TAXONOMY
Genera *Africocypha* (3 species), *Chlorocypha* (28), *Platycypha* (14), and *Stenocypha* (5)

ABOVE LEFT | Rivals of this male Beautiful Jewel (*Platycypha rufitibia*) from Gabon are threatened with the legs' red outer sides, while mates are appeased by the white insides.

ABOVE | Males of the Polychrome Jewel (*Africocypha varicolor*) from Gabon are uniquely varied, having a red, yellow, or blue tip.

JEWELS OF SOUTH ASIA TO MELANESIA

Unsurprisingly perhaps, most of the chlorocyphids' names reflect their prominent snout. The genus *Cyrano*, for example, refers to a seventeenth-century French poet's legendary nose. *Rhinocypha* means "nose-hump" in Greek. Translated literally, *Aristocypha* and *Heliocypha* have "best" and "sunny humps." *Chlorocypha* means "green-hump," although few species have any green at all. As with *themis* (p. 39) and *cnemis* (p. 167), the ending first used for *Rhinocypha* in 1842 has simply come to refer to the family.

The groups of species that many of these names currently define also appear a little inane. Most jewels found from Asia to Melanesia have narrower,

TOP LEFT | A male of the Peacock Jewel (*Aristocypha fenestrella*), widespread from Peninsular Malaysia to northern India.

LOWER LEFT | A male Philippine Jewel (*Rhinocypha colorata*), which is common from Luzon to Mindanao.

OPPOSITE | With countless marvels such as this widespread Sapphire Jewel (*Heliocypha perforata*), the Asian tropics hold the real trove of jewels.

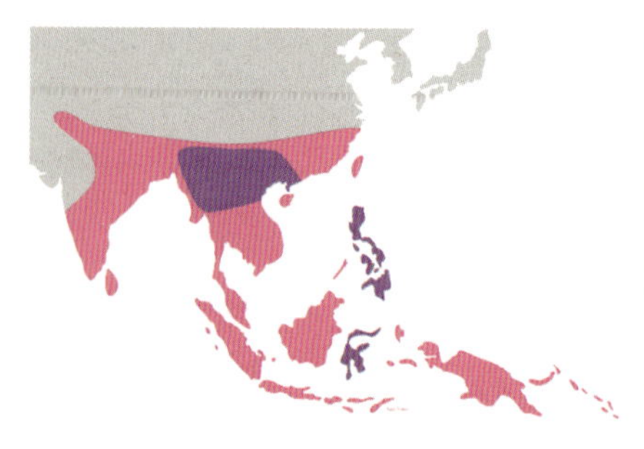

DIVERSITY
84 species of mostly forested streams and rivers from south Asia to Melanesia

TAXONOMY
Poorly delimited genera *Aristocypha* (12 species), *Calocypha* (1), *Heliocypha* (9), *Heterocypha* (1), *Rhinocypha* (46), *Rhinoneura* (2), and *Sundacypha* (2) may form subfamily Rhinocyphinae (**pink and purple on map**); *Cyrano* (2), *Disparocypha* (1), and *Indocypha* (7) might represent three further subfamilies (**purple only**), while *Melanocypha* (1) seems nearest Libellagininae (next profile), but there are too many unresolved issues for a definitive classification

less colorful abdomens than their African counterparts (previous profile), often marked extensively with black. Their wings, by contrast, commonly have dark patches that may sparkle green, blue, purple, pink, and almost any other color. Many species also have clear (equally iridescent) windows in these dark patches, as well as a brightly colored wedge on the front of the thorax. While some of the latter, found mostly on the mainland, are placed in *Aristocypha*, *Heliocypha*, and *Heterocypha*, others reside in *Rhinocypha*.

Many other species placed in the huge genus *Rhinocypha*, which range mainly across the many islands from the Malay Peninsula to the Solomons, do not have the wedge and windows. These are also absent in the genera *Sundacypha* from Sundaland and *Rhinoneura* from Borneo's mountains. The latter is also the only jewel genus with a long abdomen, although, unlike regular damselflies, the final three segments are slightly wavy.

Genetics suggest that these two groups of species—the latter more insular and the former more continental—evolved separately. Reclassifying the 74 species involved accordingly will be a major puzzle. South India's Myristica Sapphire (*Calocypha laidlawi*), for example, seems more like some species from New Guinea than those of the closer Himalayas.

Even definite proof that all these species can be grouped together as the subfamily Rhinocyphinae is still wanting. Several genera found alongside, however, are clearly distinct. With their broad bodies, bold coloration, and clear wings, the Philippine genus *Cyrano* recalls the African *Chlorocypha*, for example, as does *Indocypha* from south China and adjacent Indochina. Sulawesi's *Disparocypha biedermanni* is unusually long and narrow-bodied, with odd stigmas expanding toward the wingtips. Perhaps its build is fine-tuned to dance only for rival males. Females uniquely oviposit on vertical, mossy surfaces well above the water.

While unique with its double-banded wings, Sumatra's *Melanocypha snellemanni* most recalls some *Libellago* species (next profile), similarly missing its forewing stigmas.

GEMS

Gems (*Libellago*) are distinctly smaller than the jewels found alongside them (previous profile), measuring about ¾ in (2 cm) rather than 1–1⅛ in (2.5–3 cm). Males, moreover, typically have fewer markings on the wings, but more color on the body. To assert their dominance, many fly in front of their opponent using only their hind wings, holding their dark forewing tips outstretched and completely still. Other species, such as the dark-bodied Clearwing Gem (*L. hyalina*), do little in terms of display despite being territorial.

A few species have lost the stigmas in their dark wingtips, one from Sulawesi having the forewing's leading edge distinctly marked and thickened near the node instead. Such modifications must affect the wing's weight balance and thus be linked to distinct display behavior. Unlike most chlorocyphids, for example, these males do not woo females (and confront rivals) where they oviposit, but wherever encounters occur. On account of its sclerotized vein, this particular species was separated from *Libellago* as *Sclerocypha bisignata*. Other species on Sulawesi have a slight thickening too, however, including another placed in its own genus, *Watuwila vervoorti*. Probably, all this variation evolved from within the community of 11 gem species and subspecies unique to the island.

DIVERSITY
28 species of mostly forested streams and rivers from Sri Lanka (the only chlorocyphids there) to Sulawesi and the Lesser Sundas

TAXONOMY
The single species currently in each of *Pachycypha*, *Sclerocypha*, and *Watuwila* may eventually be placed within the genus *Libellago* (25 species)

TOP LEFT | The male Red-veined Gem (*Sclerocypha bisignata*) from Sulawesi with the peculiarly "inflamed" leading edge of its forewing.

LOWER LEFT | A male Rainbow Gem (*Libellago asclepiades*) from Sulawesi, Indonesia.

Apparently, while the major lineages within Chlorocyphidae differ substantially in wing decoration (see previous profiles), such features also evolve rapidly when many individuals of multiple species compete for space and posterity along rainforest streams. Much study of their genetics and outrageous choreographies will be needed before we fully understand how there can be over 160 species gracing Africa's and Asia's watercourses with their color and cheer today. The endangered Peatswamp Gem (*Pachycypha aurea*), for example, is only known from scattered records from lowland swamp forest on Borneo. While chubby (its name means "golden thick-hump"), it is one of the tiniest odonates at ⅔ in (18 mm). Except for clear wingtips, males are almost completely deep yellow, while females have largely brown wings with gray-pruinose bodies. How might they possibly behave?

ABOVE | A male of southern Asia's Golden Gem (*Libellago lineata*) trying to get the attention of several egg-laying females.

FAMILY: MEGAPODAGRIONIDAE
LONG-LEGGED FLATWINGS OR SPINDLESHANKS

As foregoing accounts show, both the showy damsels with their dense mesh of wing veins and the smaller ones with simpler venation usually rest with folded wings. Many less common and conspicuous groups have intermediate configurations of veins, however, and often rest with wings outstretched, allowing for quick takeoff. While this can increase their vulnerability to birds and dragonflies, such predators might be less of a threat in their habitats, typically forest streams (compare pp. 239 and 242).

Seeming alike merely by looking unfamiliar, over 300 of such species were left in two taxonomic "dustbins" for decades: Amphipterygidae (p. 209) and, especially, Megapodagrionidae. Former members of the latter are still often called flatwings or megapods colloquially, but they are placed in 15 different families today. Although genetics gave this

LEFT | A male of the Amazonian Flatwing or Spindleshank (*Megapodagrion megalopus*), whose scientific name can be translated as "large-legged damselfly with very large legs."

RIGHT | A male of Croizat's Flatwing (*Teinopodagrion croizati*) showing the exceptionally long legs of the family; as so many unrelated damselflies are called "flatwings" perhaps "spindleshanks" is a better common name.

DIVERSITY
29 species of forest streams in the tropics of South America

ADULT HABIT
Small to medium-sized damselflies with quite densely veined wings held open at rest and longer and thinner legs than similar damsels in range

TAXONOMY
Genera *Allopodagrion* (3 species), *Megapodagrion* (1), and *Teinopodagrion* (25)

reclassification its final push, clues that many were not closely related always existed in their genitalia and particularly the nymphs' gills.

Like most flatwings, the species retained in Megapodagrionidae typically perch by forest streams with the wings held flat (i.e., open) and abdomen level. While many other "megapods"—which, in fact, belong to other families—have long legs, however, these are exceptional among true megapods, as stressed twice in the Amazonian Flatwing's (*Megapodagrion megalopus*) scientific name. The nymphs have very long legs too, as well as stretched out gills and antennae, thus recalling Calopterygidae (p. 180) and Dicteriadidae (p. 224). They also have distinctive horns on the back corners of their heads.

The sole *Megapodagrion* (**pink on map**) is rather dull brown, but most other species (**blue on map**) are black with contrasting pale thorax markings. Often hunched on leaves, with the body held close to the surface and legs spread like a spider, *Teinopodagrion* occurs only along the Andes from Venezuela to Argentina, in montane cloud forests down to the foothills. *Allopodagrion* is confined to the Atlantic forest of southeastern Brazil and adjacent Argentina. As members of this family are localized, more species undoubtedly await description.

FAMILY: ARGIOLESTIDAE
PALEOTROPICAL FLATWINGS

ABOVE | This female *Caledopteryx maculata* is one of six argiolestid species restricted to New Caledonia.

OPPOSITE | Most of Australia's flatwings are not colorful, but this male Golden Flatwing (*Austroargiolestes chrysoides*) is a notable exception.

The name *Argiolestes* combines *Argia* (p. 152) with *Lestes* (p. 238), linking two vastly different families (Coenagrionidae and Lestidae) with each other. This perfectly expresses the taxonomic ambiguity with which Megapodagrionidae was long treated (previous profile). Appropriately, Argiolestidae is the biggest chunk chipped off that once massive family.

Even now the relationships are better understood, finding details that characterize all species of this (and many other) groups is challenging, and so the family is often best identified by the exclusion of others. This is largely because the obvious

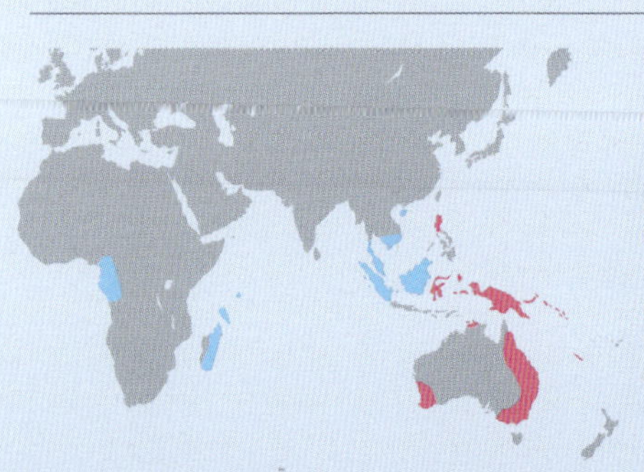

DIVERSITY
127 species of mostly running and forested waters in the tropics from Africa to Australia

ADULT HABIT
Generally fairly large damselflies, which are the dominant group within much of their range that perches by running waters with wings held open and abdomen level

TAXONOMY
Subfamily Argiolestinae (**pink on map**) with genera *Archiargiolestes* (3 species), *Argiolestes* (14), *Austroargiolestes* (10), *Caledargiolestes* (2), *Caledopteryx* (2), *Celebargiolestes* (4), *Eoargiolestes* (1), *Griseargiolestes* (7), *Luzonargiolestes* (2), *Metagrion* (23), *Miniargiolestes* (1), *Podopteryx* (3), *Pyrrhargiolestes* (7),

features such as body color and wing shape evolve very rapidly (adults only exist to reproduce and disperse, after all) and yet can vary only within strict constraints. Closely related species may thus end up looking very different, while distant relatives are alike purely due to convergence or coincidence.

Argiolestid nymphs, by contrast, are almost unique, with their horizontal fan-shaped terminal gills. Males, moreover, lack the spines on their penis's shaft that other "megapods" almost always possess. Also, while most argiolestids similarly breed in forested running waters, males perching inactively nearby with spread wings, they are more varied ecologically than other former Megapodagrionidae.

This applies particularly to the Australasian species, placed in the subfamily Argiolestinae. Australia notably harbors the only temperate flatwings, some inhabiting boggy mountain seeps and being distinctly metallic or extensively gray-pruinose, presumably to thermoregulate. Others live in waters that dry to pools in summer. *Austroargiolestes* and *Griseargiolestes* are confined to Australia's east, the notably small species of *Archiargiolestes* and *Miniargiolestes* to the southwest.

Caledopteryx and *Eoargiolestes* are among four closely related genera restricted to New Caledonia, which also include the only argiolestids to regularly close their wings (*Trineuragrion*) and that may be terrestrial, the nymph of one *Caledargiolestes* species having been found among leaf-litter along forest streams.

Solomonargiolestes (2), *Trineuragrion* (1), and *Wahnesia* (17); and Podolestinae (**blue on map**) with *Allolestes* (1), *Nesolestes* (16), *Neurolestes* (2), and *Podolestes* (9)

LEFT | Treehole flatwings such as this female *Podopteryx selysi* from Australia may be encountered randomly perched on the forest floor.

TOP RIGHT | The female of an unidentified islander (*Nesolestes*) species from Madagascar showing her abnormally elongated ovipositor

MIDDLE RIGHT | Male of the exquisite Guinea Flatwing (*Neurolestes trinervis*) from Gabon.

LOWER RIGHT | A mating pair of the delightful Blue-spotted Phantom (*Podolestes orientalis*), found in swamp forests in southeast Asia.

With Luzon's *Luzonargiolestes* and Sulawesi's *Celebargiolestes*, they represent the few groups that managed to expand beyond New Guinea and neighboring islands.

While many argiolestids are dark and dull, those in this center of the subfamily's diversity are often colorful. *Pyrrhargiolestes* species have slightly enlarged orange to red legs, *Solomonargiolestes* bold blue or orange thoracic patterns, and *Argiolestes* bright white claspers. The abdomen tip is soft on the upperside in *Metagrion* and *Wahnesia* males, probably allowing them to expand it in display.

New Guinea and adjacent northern Australia also harbor the only non-coenagrionid damselflies that breed in water-filled cavities, the treehole flatwings (*Podopteryx*). Like unrelated tree-dwellers, such as helicopters (p. 156), the inch-long nymphs have shortened and widened gills, and the huge adults (over 2⅓ in/6 cm long) hang rather than perch upright; oddly, their penises also similarly consist of a long and slender whip!

The remaining quarter of argiolestids, the subfamily Podolestinae, range from tropical Africa to Asia; most species known are islanders (*Nesolestes*) that inhabit Madagascar, with one found nearby in the Comoros, and their number may yet double (see p. 231). Their nearest relatives are localized in Africa: the Guinea Flatwing (*Neurolestes trinervis*) in the lowland rainforests around the Gulf of Guinea and the Nigerian Flatwing (*N. nigeriensis*) in just a few montane forest fragments on the border with

Cameroon. The elusive phantoms (*Podolestes*) are largely limited to Borneo, Sumatra, the Malay Peninsula, and Hainan, but may be more widespread. The distinctive Seychelles Islander (*Allolestes maclachlani*), survives only on the granitic islands there (p. 165).

The great geographic and genetic distances between the two subfamilies and their genera suggest that Argiolestidae originated when Africa and Australia were connected by Antarctica (compare p. 57 and 171), but have evolved separately for many millions of years. Ecologically the groups remain similar, however, although podolestines generally have longer ovipositors with fewer, smaller teeth. Females might thus lay their eggs in softer substrates, but oviposition in leaf stalks has been reported too. *Podolestes* inhabit sluggish runnels and leaf-littered pools in lowland swamp forest, the other genera mostly shaded seeps and streams, and *Nesolestes* open forested rivers too.

FAMILY: PHILOSINIDAE

ASHTAILS AND SIGNALTAILS

While also languishing in the "trashcan" Megapodagrionidae for ages (p. 194), philosinids are rather different creatures, more like their actual relatives, Philogangidae (next profile). Male signaltails (*Rhinagrion*) perch conspicuously fairly high on sunlit vegetation over water, wings and abdomen held (almost) horizontally. Ranging from Indochina to Java and the Philippines, the species appear to favor slower stretches of lowland forest streams, laying their eggs in muddy banks. The nymphs hide among detritus in calm spots and have unique terminal gills, the outer two folded around the median gill to form a tube.

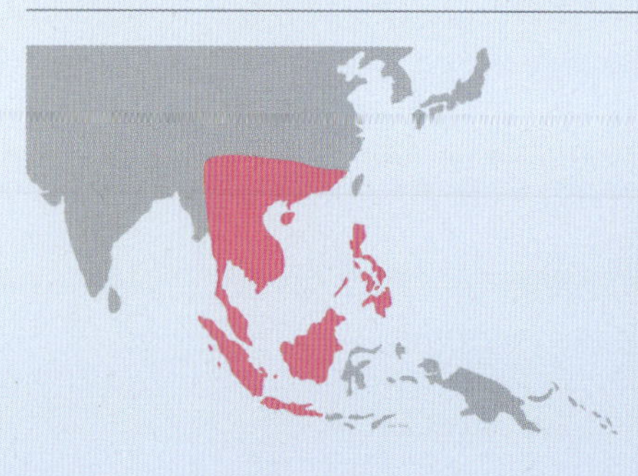

DIVERSITY
13 species of mostly open or sluggish streams and rivers, usually in lowlands, from south China to Java and Indochina

ADULT HABIT
Rather large and notably thickset with quite densely veined wings held open at rest, separated from most similar damsels within their range by their prominent faces and bright male colors

TAXONOMY
Genera *Philosina* (3 species) and *Rhinagrion* (10)

With bold patterns of blue, yellow, orange, red, and even pink, adult males are among the most striking odonates. Like many damselflies, the abdomen tip is typically brightly colored on top, but the underside can also be red, violet, or white. The males can instantly widen these segments, moreover, flaunting the colored signals on their tails (hence the common name) to optimal effect.

Although these features may correspond to displays as complex as those of jewels (p. 188), such behavior is rarely seen. Bornean Signaltail (*R. borneense*) males confront each other face-to-face, flexing their tail-ends up and down to flash the color on both sides. Other species fly side-by-side, or chase each other in circles, quickly gaining height. Flying in front of a perched female, the Chinese Signaltail (*R. hainanense*) male quivers his abdomen tip's red underside like a flag.

Similar antics are not known from the ashtails (*Philosina*), found in southern China and nearby Vietnam and Laos. They are even more robust, but otherwise differ mainly by the mature males' dense whitish pruinosity. This covers the entire body in the signaltail-sized Plain Ashtail (*P. alba*), but spares the bright red abdomen tip of the much larger Red-tipped Ashtail (*P. buchi*).

Ashtail nymphs look similar, but *P. alba* breeds in shaded forest streams, often with rocks and considerable flow, while *P. buchi* favors open rivers, which can be over 300 ft (100 m) wide and lie in completely developed landscapes. Males perch close to the water, probably to defend territories, and fly strongly, thus recalling *Orthetrum* dragonflies (p. 36).

OPPOSITE | A male Golden-tipped Signaltail (*Rhinagrion viridatum*) from Thailand; being so colorful and bold, philosinids have also been called rainbow flatwings.

ABOVE RIGHT | A male Red-tipped Ashtail (*Philosina buchi*) from southern China.

RIGHT | The male Sabah Signaltail (*Rhinagrion elopurae*), known only from northeastern Borneo.

FAMILY: PHILOGANGIDAE

TITANS

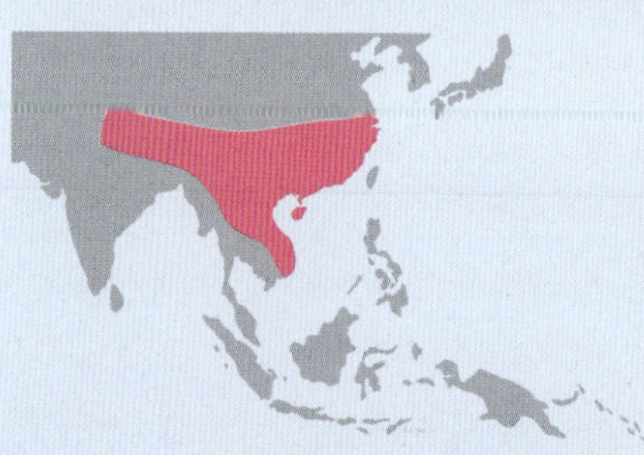

DIVERSITY
4 species of fast-flowing wooded streams from mainland Southeast Asia

ADULT HABIT
Robust damselflies with prominent faces as well as long, narrow wings held open at rest; best recognized within range by their great size (about 3 in/7–8 cm long) and dense venation

TAXONOMY
Genus *Philoganga* (4 species)

As their common name indicates, titans are among the biggest and heaviest damselflies. Indeed, with their bulk and outstretched wings, they are easily mistaken for dragonflies! Like other damsels with slender but densely veined wings, the four very similar species were long treated as amphipterygids (p. 209). Genomics, however, revealed that Philosinidae not only look alike superficially and have a similar range, but are their closest relatives (previous profile).

The most notable similarity is their prominent muzzle, which gave *Rhinagrion* ("nose-damsels") their scientific name. This corresponds with an unusually narrow labium (lower lip) and long mandibles, the former deeply cleft, so perhaps they feed differently from other damselflies. Other shared features are the long stigmas, short leg spines, simple pincerlike male claspers, and of course their robust build.

The sturdy nymphs are also similar, with a median terminal gill flanked by longer outer gills. The latter are not tubelike, however, all three gills being swollen like balloons and narrowing gradually into a fine point. Rather than perch at the water, adult males hang (almost) vertically in vegetation, often high up and meters away, only occasionally making quick patrols over the stream. Females are most often seen far from the water.

OPPOSITE | An old dull-colored male Ochre Titan (*Philoganga vetusta*) from Vietnam.

ABOVE | A fresh male Ochre Titan (*Philoganga vetusta*) from Hong Kong.

FAMILY: EUPHAEIDAE
SATINWINGS

While nymphs of mayflies—odonates' nearest relatives—all respire with lateral gills, just two damselfly families possess similar structures, a fringe of fingerlike appendages low on the abdomen sides (see photo p. 9). These are protected by their broad, flattened bodies as they cling to the underside of rocks or coarse detritus in fast-flowing water with sturdy legs directed sideways. The nymphs (and those of many families treated next) recall stoneflies too, and they may fill similar niches, stoneflies being represented poorly in the tropics. The three terminal appendages, used as gills by other damsels, may have a more sensory function, being swollen like hairy balloons and ending in a long filament.

Large with densely veined, often boldly marked wings, the adults recall the calopterygids found alongside them (p. 180), although the nymphs'

LEFT | A male Malayan Black Velvetwing (*Dysphaea dimidiata*) from Malaysia in its characteristic dragonfly-like posture.

OPPOSITE | A male Ochraceous Satinwing (*Euphaea ochracea*) from Malaysia.

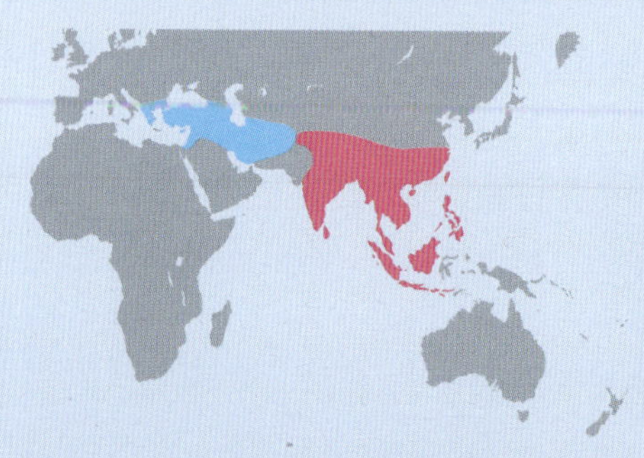

DIVERSITY
79 species found at all kinds of running waters from southeastern Europe to the Philippines and Lesser Sundas

ADULT HABIT
Mostly large damselflies recalling calopterygids due to size and densely veined, often boldly marked wings, but generally more robust with shorter legs, long stigmas, and mostly dark non-metallic bodies, sometimes marked with bright or pastel colors

TAXONOMY
Genera *Anisopleura* (11 species), *Bayadera* (17), *Cryptophaea* (3), *Cyclophaea* (1), *Dysphaea* (9), *Epallage* (1; **blue on map**), *Euphaea* (35), *Heterophaea* (1), and *Schmidtiphaea* (1)

sturdiness is reflected in their robust build and short legs. Male satinwings (*Euphaea*) are especially colorful, with deep amber or (more often) extensively black wings, typically with strong green, blue, purple, or red reflections. Their bodies are mostly black, but some species have bright blue thorax sides or are glowing red. Euphaeids have much longer stigmas than calopterygids in their usually narrower wings, and (while often perching with closed wings too) an almost dragonfly-like posture. Europe's only euphaeid, the Odalisque (*Epallage fatime*), sits at the tip of a perch with abdomen and spread wings slightly raised. Perching in the sun along tropical Asia's lowland rivers, the largely black male velvetwings (*Dysphaea*) may even press their open wings down, just as many libellulids do (see p. 54). They and *Euphaea* males possess "pseudoauricles," triangular projections on the sides of the abdomen base, which must have a similar function to auricles in dragonflies (p. 74). These are drawn out into downward spurs in Palawan's Gunslinger (*Cyclophaea cyanifrons*), like holsters on the hips. Luzon's wholly red Bearded Satinwing (*Heterophaea barbata*)—the largest euphaeid at up to 2¾ in (7 cm) long—is another clear-winged but spectacular genus unique to the Philippines.

The males of these showy genera perch on rocks and logs along streams and rivers, sometimes in high densities. While aggressive, they do not perform elaborate displays (compare calopterygids and chlorocyphids; p. 188). Potential territories might not be patchy enough to defend, perhaps because nymphs can live under almost any object in the current. Females appear to descend as deeply as possible into fast-flowing water to lay their eggs, sometimes braving the spray of waterfalls, and often accompanied by their mate. Benefiting from their heavy build, some females even dive straight into the torrent.

Occurring on slopes and in foothills from the western Himalayas to southern China, the velveteens (*Bayadera*) and mountaindarts (*Anisopleura*) are less showy. The former recall *Epallage*, with often entirely blue-pruinose males with darkened wingtips. While they perch quite visibly along rocky streams and rivers, the almost clear-winged *Anisopleura* sit lower by steep seeps and streamlets, their dark bodies marked subtly with white pruinosity and pastel yellows, greens, and blues.

The male hind wing's leading edge is often expanded near its base, a unique, unexplained feature.

Four small clear-winged species placed in *Cryptophaea* and *Schmidtiphaea*, finally, are so dainty and long-bodied that they may not recall euphaeids at all. Known from a few shadowy streams on forested mountain slopes from southwest China to northeast India, they seem nearest *Bayadera*.

Altogether, euphaeids seem most like the American polythorids, whose nymphs also have lateral gills (p. 218). Their nearest relatives, however, are the strange, isolated families from Australia, New Guinea, and Hainan, treated next.

ABOVE LEFT | The female *Cryptophaea vietnamensis* may not be recognized as a euphaeid instantly, her orange (sometimes purplish) thorax recalling some chlorocyphids and polythorids.

BELOW | A pair of Odalisques (*Epallage fatime*) in tandem in Turkey.

FAMILY: LESTOIDEIDAE

BLUESTREAKS AND ROCKMASTERS

Found only along rainforest streams in northeastern Australia, the bluestreaks (*Lestoidea*) are rather inconspicuous and normal-looking damselflies, small and largely black with yellow to blue sides (hence their common name), keeping their clear, narrow wings closed at rest (see photo p. 5). The bulky rockmasters (*Diphlebia*) could hardly differ more, seeming more like dragonflies. The black-and-blue males fly rapidly from rock to rock, perching with the often black-banded wings (one species has narrow white bands) spread out above their muscular thorax. Females are generally much duller, often extensively brown.

The latter genus inhabits streams and rivers in eastern Australia, two of the five species tolerating waters that intermittently dry up into pools. Of these, the Tropical Rockmaster (*D. euphoeoides*) extends into southern New Guinea. Living under stones and among detritus, the flattened nymphs recall those of stoneflies (p. 204). They have big square heads and three swollen terminal appendages ending in a fine point. Although the remarkable similarity of *Lestoidea* nymphs was known for decades, genetic research only recently confirmed how closely related the two Australian genera are. This work also revealed that their next nearest relatives are Euphaeidae: no surprise for the showy rockmasters, but the bluestreaks too do not look so different from more modest euphaeids.

LEFT | The impressive male of the Sapphire Rockmaster (*Diphlebia coerulescens*).

DIVERSITY
9 species of streams and rivers in eastern Australia and southern New Guinea

ADULT HABIT
Medium-sized to very large, generally robust damselflies with comparatively long stigmas, *Lestoidea* differs from other damsels in range by combining this with closed wings at rest, while *Diphlebia* is exceptionally thickset with densely veined, often marked wings held open

TAXONOMY
Genus *Diphlebia* (5 species) in subfamily Diphlebiinae and *Lestoidea* (4) in Lestoideinae; sometimes treated as separate families

FAMILY: PSEUDOLESTIDAE
PHOENIXES

Named for Fenghuang, the birdlike chimera of Chinese mythology, the Phoenix (*Pseudolestes mirabilis*) is one of the strangest and most spectacular odonates. Its hind wings are much shorter than the forewings, looking shriveled with narrow cells and an odd rhomboidal shape. In males, waxy fibers on the underside reflect light more strongly than ordinary pruinosity, creating a silver-white brilliance below and, cast through the amber-tinted membrane, patches of glowing copper on the upperside.

The males perch horizontally, angling their open wings to catch the light. Territories barely a meter wide are often densely packed, so males clash constantly, flying with their bright blue faces toward each other, hind wings held down and completely still. Most contests last a few minutes, but some an hour. By contrast, males rarely encounter females and yet spare only seconds to court them. They transfer sperm to the secondary genitalia after copulation, rather than before like other odonates, perhaps to always be ready to mate (p. 17).

Hiding among gravel and stones, the nymph is unique too: three large spherical appendages shield tufts of retractable gills on the abdomen tip's underside. Such tufts are otherwise present only in four distantly related genera, now each considered separate familes (see opposite), but genomics suggests that the Phoenix's nearest relatives are the fairly nearby Lestoideidae and Euphaeidae (previous pages).

LEFT | While the world range of the Phoenix (*Pseudolestes mirabilis*) measures less than 55 by 110 miles (90 by 180 km), its densities are high, so this unique family is not considered under threat.

DIVERSITY
Single species confined to mostly rocky, forested upland streams on Chinese island of Hainan

ADULT HABIT
Medium-sized (about 1¾ in/4.5 cm long) and yet quite robust damselflies with uniquely shortened and widened black-and-copper hind wings that perched males spread and angle back- and upward

TAXONOMY
Genus *Pseudolestes* (1 species)

FAMILY: AMPHIPTERYGIDAE
MAYAN DAMSELS

Nymphs of the Mayan damsels (*Amphipteryx*) inhabit coarse gravel and leaf-litter in faster-flowing sections of smaller running waters at 2,300–5,900 ft (700–1800 m) elevation in the cordilleras of northern Middle America. Their adults perch nearby on overhanging vegetation.

Also having narrow but densely veined wings, the equally localized *Devadatta* and *Philoganga* from tropical Asia (pp. 202 and 210), *Diphlebia* from Australia (p. 207), *Pentaphlebia* from Africa (p. 229), and *Rimanella* from northern South America (p. 228) were long placed in Amphipterygidae too. Except for *Philoganga* and *Diphlebia*, these genera also share retractable gill tufts implanted near the base of the nymph's terminal appendages, a peculiar feature found otherwise only in Pseudolestidae (see opposite). The tufts are probably the main respiratory organs and allow the appendages to assume other (unknown) functions, thus differing strongly in shape among the genera. They are long and conical in *Amphipteryx*, swollen at the base and tapering to a point, and densely covered with scalelike hairs.

While genetic studies prompted the placement of each genus in its own family, *Devadatta* is likely the nearest living relative of *Amphipteryx*. Structurally and ecologically they are quite similar, although *Amphipteryx* is much more colorful, males being black marked brightly with yellow, green, and blue on the face, thorax, and abdomen tip.

LEFT | A male Mexican Maya (*Amphipteryx agrioides*).

DIVERSITY
5 species of seeps and small streams, mostly in cloud forest from southern Mexico to Honduras

ADULT HABIT
Fairly large, robust damselflies that are unique within their small range due to densely veined, clear, distinctly stalked (bases narrow and largely without cross-veins) wings with extremely skewed stigmas (barely touching wings' leading edge) that are closed at rest

TAXONOMY
Genus *Amphipteryx* (5 species)

FAMILY: DEVADATTIDAE
GRISETTES

Despite their scientific name meaning "god given" in Sanskrit, grisettes (*Devadatta*) are humble creatures that can be hard to spot, typically perching motionless close to the ground in deep forest shade, usually near their breeding sites. Although one species has a bright blue face and two others blue-gray and golden yellow thorax sides, adults are largely dark with indistinct pale markings and perhaps a light cast of pruinosity.

RIGHT | A male Malayan Grisette (*Devadatta argyoides*).

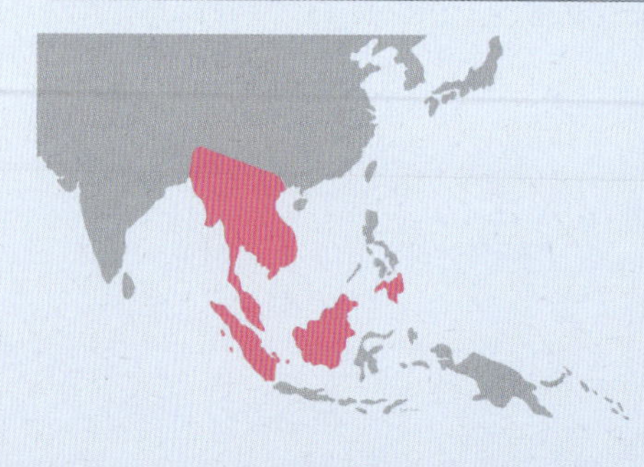

DIVERSITY
14 species of deeply shaded streamlets, mostly in hilly and mountainous terrain, from south China through Indochina to Sumatra and the Philippines

ADULT HABIT
Fairly large, dull damselflies, which may be mistaken for inconspicuous euphaeids or calopterygids due to their densely veined and often dark-tipped wings that are closed at rest; these differ in having very distinct stalks (long narrow base) and rhomboid stigmas (notably acute inner angle)

TAXONOMY
Genus *Devadatta* (14 species)

The stout nymphs live among leaf-litter or rootlets and under stones in trickles and small streams. Like other "amphipterygid" nymphs (p. 209), they have peculiar morphology, including retractable gill tufts and three enlarged pyramidal terminal appendages.

Being so cryptic, more than the 13 species currently recognized may actually exist. Careful analysis of genetics and morphology, for example, revealed that Borneo's single species actually consisted of at least five, with often two and sometimes three occurring on the same stream.

FAMILY: THAUMATONEURIDAE
CASCADE DAMSELS

While most nymphs cannot withstand the current in waterfalls, their spray creates habitat for many specialized damselflies, such as the Platystictidae (p. 233). Most spectacular may be the Great Cascade Damsel (*Thaumatoneura inopinata*) from submontane rainforests in Costa Rica and Panama (blue on map); at almost 3⅛ in (8 cm) long, it is among the New World's heaviest damselflies. The wings are greatly expanded, especially their middle, where males bear a broad black band. Females have extensively black wingtips instead, while both sexes can be clear-winged too. Males hang down from sunny branches and lianas, occasionally flashing their closed wings open in display. They fly up and down the face of waterfalls or back and forth along vertical banks of seeping water.

Females lay their eggs in the wet moss, roots, and mud covering the rocks, while the nymphs live in the watery film. They are squat with sturdy legs and three short terminal appendages,

DIVERSITY
16 species of mostly small, forested mountain streams, seeps, and wet rockfaces from southern Mexico to Panama

ADULT HABIT
Medium-sized to very large damselflies with densely veined, often distinctly marked wings that are closed at rest; might be confused with distantly related but overlapping polythorids, but wing bases have notably fewer cross-veins

TAXONOMY
Genera *Paraphlebia* (15 species) and *Thaumatoneura* (1)

which are swollen at their base but strongly constricted in the middle, end in a thick filament, and are covered with stiff hairs.

Found from Nicaragua northward (pink on map), lesser cascade damsels (*Paraphlebia*) have similar nymphs and habits, but may occupy seeps and small streams too. The nymphs can thus also be found under stones or among fallen leaves or vegetation, and are probably able to survive for several days out of water in moist leaf-litter.

OPPOSITE | The Great Cascade Damsel (*Thaumatoneura inopinata*) from the submontane rainforests of Costa Rica and Panama is among the New World's heaviest damselflies.

BELOW | The Showy Cascade Damsel (*Paraphlebia zoe*) from Mexico is one of five *Paraphlebia* species with dark-tipped males, but the only one with a white band too.

Paraphlebia adults are not as large, though, with normally proportioned and less densely veined wings. Eleven of the 15 species known were only named in 2022, following targeted fieldwork, as the dark-bodied adults are easily missed, sitting still in deep shade, often well beyond reach in steep terrain. Many species are restricted to small mountain areas, moreover, often together with *Amphipteryx* (p. 209).

Five *Paraphlebia* species have males with dark wingtips and, like *Thaumatoneura*, each appears to have a clear-winged form too. Such males can be smaller and, while they may hold territories to attract females, just as the fancy-winged males do, they often move around and try to intercept incoming females instead (compare p. 182).

Genomic research indicates that this family and the next, both formerly included in Megapodagrionidae (p. 194), are the nearest surviving relatives of the previous two families. Interestingly, each pair consists of one family confined to Middle America and another restricted to Southeast Asia.

FAMILY: RHIPIDOLESTIDAE

SPINETAILS, SHADESEEKERS, AND PHANTASMS

This Asian family, yet another small assemblage of former "megapods," (p. 194) is related most closely to the Thaumatoneuridae from Middle America (previous profile). That geographic pattern, mirrored by their next nearest relatives, Devadattidae (p. 210) and Amphipterygidae (p. 209), suggests an origin on the supercontinent of Laurasia (pp. 105 and 111), which had fractured into the present continents by about 60 mya.

The four are indeed quite similar, with swollen and hairy terminal appendages in nymphs, for example, and often marked wings with dense venation toward their margins. Although the colorful *Amphipteryx* and black-winged

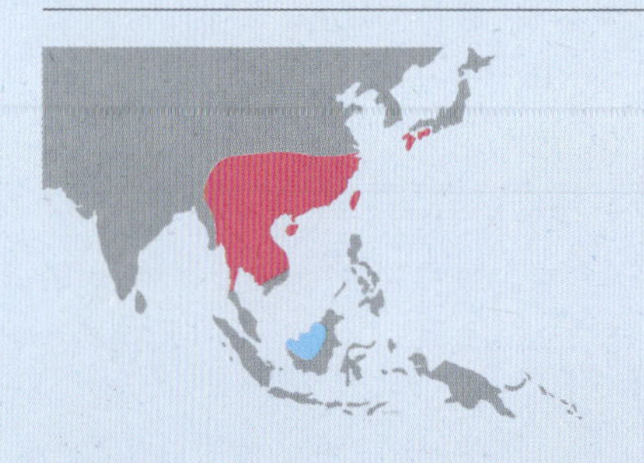

DIVERSITY
31 species of mostly very small and shady streams from Indochina to southern Japan and Borneo

ADULT HABIT
Generally medium-sized damselflies with fairly dense venation and rhomboid stigmas in their wings; typically brightly colored accents on the male's body, notably on face and legs

TAXONOMY
Genera *Agriomorpha* (2 species), *Bornargiolestes* (3), *Burmargiolestes* (3), and *Rhipidolestes* (23)

Thaumatoneura stand out more, the adults are generally rather dark damsels, with at most some color on the face or thorax. They perch with closed wings and a (nearly) level abdomen in deep shade by the smallest running waters, often very locally. Shadeseekers (*Bornargiolestes*), for example, are known from very few individuals in steep, forested terrain on Borneo (**blue on map**), breeding only where the barest amounts of water run above (or even just below) the leaf-litter, such as at dripping cliffs or tiny trickles and seeps by small streams.

OPPOSITE | Males of many rhipidolestids, such as this still unnamed spinetail *(Rhipidolestes)* from China, have brightly colored faces and legs.

ABOVE | Most spinetails *(Rhipidolestes)* have very small ranges, *R. truncatidens* being limited to south-eastern China.

The phantasmas (*Agriomorpha*, *Burmargiolestes*) from Indochina and southern China are close to them, although males often have bright yellow, orange, or blue faces.

Extending from there to Taiwan and southern Japan, spinetails (*Rhipidolestes*) are another standout group. Males bear a unique (either single- or double-pointed) prominence on top of the penultimate abdomen segment, for which they got their common name. They are the only species in the four families to perch with open wings. Comparatively varied in appearance, the wings can have dark tips, transverse bands, or yellow or red stigmas; the legs can be brilliant yellow or red; the face also blue or orange; and the head, thorax, and abdomen tip can be patterned with bright pruinosity too.

FAMILY: MESOPODAGRIONIDAE

LEAFBASKERS

When genetics first confirmed that many odd damselflies that had been classified together were not directly related (p. 194), several were initially condemned to taxonomic limbo: they did not fit the families they had been placed in, but molecular techniques were too unrefined to uncover their actual (probably ancient) affinities (p. 27). Once genomics allowed for thousands of base-pairs to be analyzed, these damsels still proved to be exceptional, however. It was not until as recently as 2022, therefore, that several new families were named.

Leafbaskers (*Mesopodagrion*) are thickset, predominantly black damselflies, which often perch flat on top of exposed leaves with wings outstretched. As they appear early, in springtime, this must help them to warm up in their mountainous home. The exceptionally stocky nymphs live among mud and detritus in the shallow water of swampy spots fed by streams, always above 3,200 ft (1,000 m) and often well over 6,500 ft (2,000 m) above sea level. Their three terminal gills are broad, pointed, and flattened, like a horizontal fan; something similar is found only in the distantly related Argiolestidae (p. 197) and Protolestidae (p. 230). The head is weirdly expanded into densely spiny "sideburns" behind the eyes.

DIVERSITY
2 species of stream-fed wet spots in mountains of China and bordering Vietnam, Thailand, and Myanmar

ADULT HABIT
Rather large, notably robust damselflies that perch flat with wings outstretched; recognized definitively by double-pointed triangular extension between male's pincerlike claspers

TAXONOMY
Genus *Mesopodagrion* (2 species)

ABOVE | A male Eastern Leafbasker (*Mesopodagrion yachowensis*).

FAMILY: AMANIPODAGRIONIDAE

AMANI FLATWINGS

The Amani Flatwing (*Amanipodagrion gilliesi*) was discovered for science by mayfly biologist Mick Gillies in 1959. Like most peculiar damselflies, it was placed in Megapodagrionidae (p. 194), its name combining *Megapodagrion* with *amani*, the Swahili word for "peace." The species is known from a single stream near the village of Amani, which runs through the remaining lower-lying forest in Tanzania's East Usambara Mountains.

While nymphs are yet to be found and described, the rather lethargic adults hide under vegetation overhanging the stream, the darkest spots in their shady habitat, not moving far when disturbed. Being quite large (just over 2 in/5 cm) and dark, and resting with wings spread and the long abdomen drooping, they recall the unrelated *Chlorolestes* (p. 245), even having white-pruinose abdomen tips and broad brown wing bands.

The remaining forests of the Eastern Arc, an old mountain chain that runs across Tanzania, harbor some of Africa's most ancient, distinctive, and threatened fauna and flora. Only in 2022 did genomics confirm that *Amanipodagrion* has no close relatives and thus deserves its own family. The IUCN Red List ranks it as Critically Endangered, the highest level for species surviving only in the wild.

ABOVE | A male Amani Flatwing (*Amanipodagrion gilliesi*).

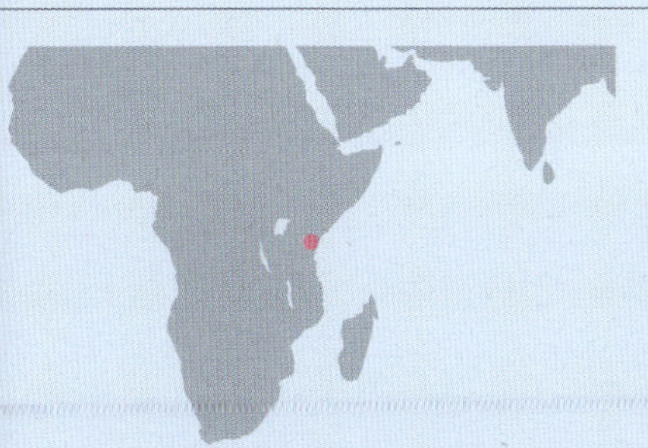

DIVERSITY
1 species known from a single rocky forest stream in northeast Tanzania

ADULT HABIT
Fairly large damselfly unique within range due to its brown-banded wings that are spread at rest

TAXONOMY
Genus *Amanipodagrion* (1 species)

FAMILY: POLYTHORIDAE
BANNERWINGS

Being rather isolated from the rest of the world, the American tropics gave rise to an impressive array of unique damselfly groups, often with peculiar nymphs. Living under stones and debris in rapid stream sections or on rockfaces wettened by cascades (but perhaps among leaf-litter in calmer spots too), nymphs of polythorids are especially sturdy with thick skin, strong legs, and spines along the abdomen's back. Swollen and adorned with several claw-like thorns, the three terminal appendages recall inflated rubber gloves.

These appendages' respiratory function is replaced by fingerlike gills on the side of all except the first and last three abdominal segments. Only Eurasia's euphaeids have lateral gills as well—also present on the third-last segment (p. 204).

TOP | A male Blue Cora (*Cora marina*) from Guatemala.

LEFT | A male Red-fronted Glitterwing (*Chalcopteryx rutilans*) from Peru flashing his wings.

OPPOSITE | A male of the Orange-banded Bannerwing (*Euthore fassli*) from Colombia.

DIVERSITY
62 species of forested running waters in the American tropics

ADULT HABIT
Generally large and sturdy damselflies with very densely veined, often elaborately marked wings that are typically closed at rest; unlike the calopterygids within range (p. 187), these are abruptly narrowed close to their base, as if they have a short stalk, and invariably bear large stigmas

TAXONOMY
Genera *Chalcopteryx* (5 species), *Chalcothore* (1), *Cora* (9), *Euthore* (14), *Miocora* (10), *Polythore* (23), and *Stenocora* (1)

With their sturdy bodies, short legs, and often colorful wings, the adults are surprisingly similar too, with broadly similar behavior. These damsels probably owe their similarities to being adapted to withstanding the current in fast-flowing waters, possibly including the impact of rainy-season spates.

Most diversity occurs along the mountainous spine from south Mexico to Bolivia. *Polythore* species have wide and densely veined wings, usually with broad and sometimes blue-iridescent black bands and often white, yellow, or orange bands too, inspiring the name bannerwings. Some *Euthore* are almost as spectacular, while others are more like *Cora* and *Miocora*. With dark bands or tips at most, sometimes some blue iridescence, and often just clear wings, these species are called coras.

As in other "Caloptera" (p. 180), males use their beautiful wings in display. Some glitterwings (*Chalcopteryx*) have largely dark hind wings, for example, that sparkle in multiple colors. As in *Pseudolestes* from Hainan (p. 208), these are much shorter and broader than the clear forewings. While the latter keep the male aloft, he flaunts the iridescent wings at potential mates. On account of its smaller size (up to 1⅓ in/3.5 cm long, rather than 1½ in/6.5 cm) and lowland distribution in Amazonia, this genus forms a distinct group within the family.

Both sexes of many fancy-winged species vary with age, or have forms matching the opposite sex.

Typical males of Costa Rica's Variable Cora (*Miocora semiopaca*) are rather large with dark-tipped wings, for example, while others are clear-winged like females. A third morph has clear wings and a largely yellow (but locally blue!) face and thorax, however, and is also warier. This form resembles the Blue Flatwing (*Philogenia peacocki*; p. 227), but why both species look like this, or why one might even imitate the other, is unknown. They only occur together very locally, while similar coloration occurs in distant but ecologically comparable euphaeids (p. 204) and devadattids (p. 210) too.

Why some *Polythore* and *Euthore* species mimic glasswing butterflies, even imitating their flight motion, is more obvious, as those are unpalatable to many birds. Strongly reflecting ultraviolet light, the white-pruinose bands are particularly important in this deception.

Much nonetheless remains poorly understood about this family. Some *Polythore* males grasp the female with their claspers by the thorax directly behind her prothorax, for example, rather than by the prothorax itself (compare p. 18). Tandems are so rarely observed, however, that it is unclear how widespread such deviant habits are.

Two genera with single species are virtually unknown. Limited to southeast Venezuela and adjacent Guyana (p. 228), the Tepui Glitterwing (*Chalcothore montgomeryi*) is near *Chalcopteryx*, but has dark basal halves of all wings. Restricted to cloud forest in the Andes of Peru and Ecuador, the Horned Cora (*Stenocora percornuta*) may be the most disparate polythorid, with its clear narrow wings and huge upright horn on the male's last abdominal segment, like a wizard's pointy hat.

OPPOSITE | The typical male form of the Variable Cora (*Miocora semiopaca*) from Costa Rica.

ABOVE | The female of the Giant Bannerwing (*Polythore gigantea*) from Colombia.

RIGHT | A small nymph of *Polythore derivata* from Peru showing its many peculiar features.

FAMILY: HETERAGRIONIDAE AND MESAGRIONIDAE

FLAMBOYANTS, REDLEGS, AND KIN

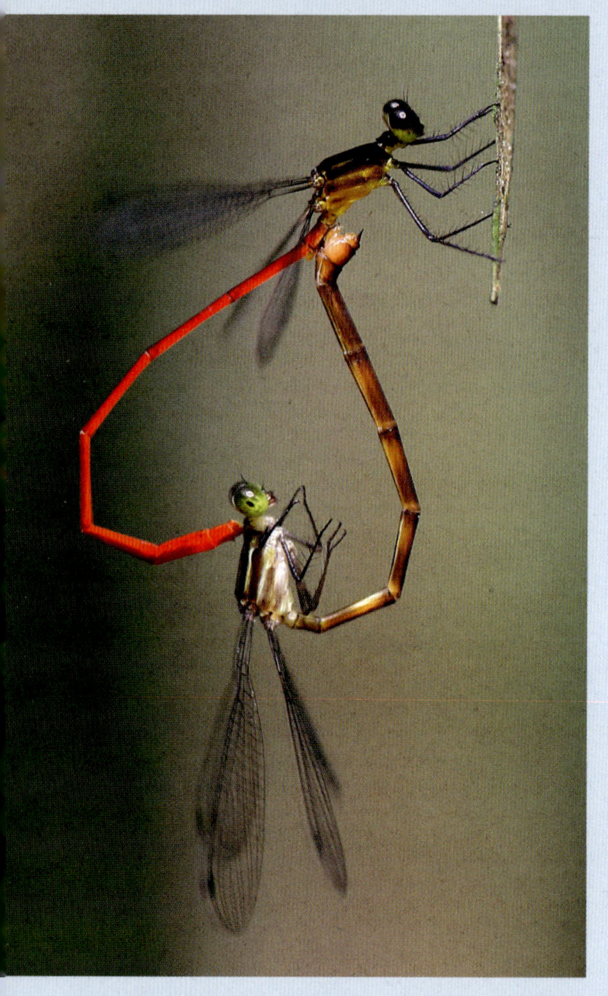

Odonates may favor sunny weather, but Amazonia's raindamsels (*Oxystigma*) appear to be most active during (even heavy) downpours. Their males are rather dark and dull, marked only with some blue. Perhaps this coloration is linked to their preferred weather, as their closest cousins are the New World's brightest flatwings.

With over 60 species named, the flamboyant flatwings (*Heteragrion*) form the largest group of former megapodagrionids from tropical Mexico to Argentina (p. 194). Males are marked on the face, thorax, and abdomen with glowing red, orange, and yellow, but sometimes with white or blue. They guard pools in streams and springs in the dim forest understory. When perched on twigs or leaves with wings spread, or hovering face-to-face to challenge each other, their long abdomens often hang down a bit.

Confined to the Guiana Shield, most tepui flatwings (*Dimeragrion*) are found on the tabletop mountains of Venezuela (p. 228). Males behave like *Heteragrion*, perching with open wings near small forest streams, but are black with bright pruinosity on thorax and abdomen tip and a contrasting white

DIVERSITY
78 species of forested running waters in the American tropics

ADULT HABIT
Mostly medium-sized damselflies diverse in appearance but typically the only forest stream damsels within range with fairly densely veined wings, a sharp ridge between the antennae (absent only in *Oxystigma*), often bright color (notably reds to yellows) on parts of the male body, rhomboid stigmas (with acute inner angle), and/or spread wings at rest

TAXONOMY
Genera *Dimeragrion* (6 species), *Heteragrion* (62), *Heteropodagrion* (5), and *Oxystigma* (3); *Mesagrion* (1) of Mesagrionidae and unassigned *Sciotropis* (1) may belong in Heteragrionidae too

EFT | Mating wheel of the Red-and-black Flamboyant (*Heteragrion erythrogastrum*) from Panama.

RIGHT | Male of the Sanguine Redleg (*Heteropodagrion sanguinipes*) from Ecuador.

BELOW | Male of the Striped Raindamsel (*Oxystigma petiolatum*) from Suriname.

or yellow labrum (upper lip). This facial marking recalls the redlegs placed in *Heteropodagrion* and *Mesagrion*, which are restricted to the Andes of Colombia and Ecuador. In all three genera the third-last abdominal segment's armor is uniquely split across the top in females.

Both Andean genera breed in dark forest on rockfaces wettened by small waterfalls. The blackish adults rest with closed wings, males having deep red on (parts of) the legs and abdomen, and sometimes on the stigmas and prothorax too, rather resembling some distantly related Rhipidolestidae (p. 214).

While genetic analysis assigned *Dimeragrion* and *Heteropodagrion* to Heteragrionidae, *Mesagrion* was placed in its own family, Mesagrionidae. More in-depth study may combine these genera as one family again, however, uniting all "megapods" with a sharp ridge between the adults' antennae, similar to that found in some coenagrionids (p. 153).

The only extant damselflies left unassigned to any family (p. 216) are the wood-elves (*Sciotropis*) from Venezuela's coastal cordilleras. While genetic study has been inconclusive, the nymphs recall those of redlegs, with large heads, strong legs, and three swollen and hairy appendages, allowing them to inhabit similar "sheet flow" on forest-shaded cliffs. Adults settle with wings outspread, however, and have some peculiar features.

FAMILY: DICTERIADIDAE

BARELEGS

Adults of virtually all of some 6,400 odonate species capture prey with their legs. Spines and bristles give these grip and feel, but also form into a mesh that can scoop victims up like a basket or be cast over them like a net. Dicteriadid limbs virtually lack ornamentation; long and fragile, like the legs of harvestmen, these seem unsuited to grabbing prey. They appear to catch insects in their mouths instead. The postclypeus (which holds the labrum or upper lip) widens toward the front,

while the palps (fingerlike structures on the labium or lower lip) are unusually long, ideal for holding their quarry. Just why a tried-and-tested technique was abandoned is unknown.

The nymphs also have thin, spidery legs, as well as three equally fine and spindly terminal appendages. They thus recall their adult stages, but also the lanky nymphs of calopterygids (p. 180) and megapodagrionids (p. 194); presumably they inhabit a similar underwater niche of tree rootlets and leaf-litter.

DIVERSITY
2 species restricted to running waters in lowland rainforest in tropical South America

ADULT HABIT
Quite large damselflies, which are unique for virtually lacking spines and bristles on their extremely slender legs

TAXONOMY
Genera *Dicterias* (1 species) and *Heliocharis* (1); the family is sometimes called Heliocharitidae

ABOVE LEFT | A male of the Red Bareleg (*Dicterias atrosanguinea*), found only at small streams along the lower Amazon River.

ABOVE | The male of the larger Blue Bareleg (*Heliocharis amazona*), widespread elsewhere in Amazonia.

FAMILY: HYPOLESTIDAE

CARIBBEAN FLATWINGS

Like many so-called flatwings (p. 194), *Hypolestes* species perch over forest streams on long egs with the wings and abdomen spread out orizontally. Largely black with a few pale lines, becoming pruinose-white on thorax and abdomen ips in mature males, they are indeed remarkably ike argiolestids from Madagascar or Australia p. 196). Genetics revealed no close relatives, owever, with Dicteriadidae as likeliest (but perhaps surprising) candidates (see opposite).

The three species are rather similar and very ocalized, the Cuban Flatwing (*H. trinitatis*) being onsidered Vulnerable to extinction on the IUCN Red List and the Jamaican Flatwing (*H. clara*) even Endangered (p. 29). Named only in 2015 and anked as Near Threatened, the Hispaniolan Flatwing (*H. hatuey*) is least at risk. Hatuey was a Taíno Cacique who traveled from Hispaniola to Cuba in the sixteenth century to fight the Spanish olonizers. Captured and burned alive, he is still enerated today. Considering how precarious the nown species' fates are, a possible fourth species n Puerto Rico, the smallest of the four Greater Antilles, might easily have disappeared before it ould be recorded there.

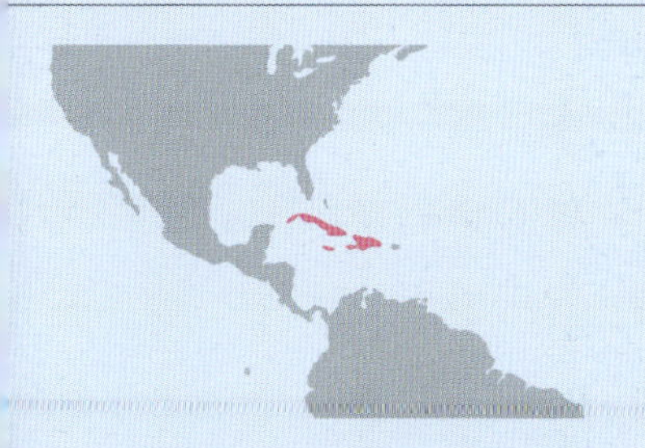

DIVERSITY
3 species of forested streams on Cuba, Jamaica, and Hispaniola

ADULT HABIT
Medium-sized damselflies unique within their range for perching with the wings and abdomen spread out horizontally

TAXONOMY
Genus *Hypolestes* (3 species)

ABOVE TOP | Male of the Hispaniolan Flatwing (*Hypolestes hatuey*).

ABOVE LOWER | Female of the Hispaniolan Flatwing (*Hypolestes hatuey*).

FAMILY: PHILOGENIIDAE
DUSKY FLATWINGS

While Heteragrionidae (p. 222) can usually be picked out by the males' bright colors, and Megapodagrionidae (p. 194) by the particularly long legs, the third major group in tropical America that was long retained in the latter family is comparatively nondescript. All *Philogenia* species are rather large (mostly 1½–2⅓ in/4–6 cm long), males often being bronzy brown on top of their thorax, but becoming darker and dusted lightly with pruinosity (densely so on the rather distinctively expanded abdomen tip), thus appearing rather dull overall. They perch inconspicuously with outspread wings and horizontal abdomen, often on the tops of leaves but close to the ground, near seeps, springs, and riffles in streams.

DIVERSITY
45 species of smaller running waters, mostly in hillside and mountain forest, from Honduras to Bolivia

ADULT HABIT
Medium-sized to large, dull-colored damselflies with quite densely veined wings held open at rest

TAXONOMY
Genera *Archaeopodagrion* (6 species) and *Philogenia* (39)

Varying primarily in the shape of the male claspers and the corresponding structures on the female prothorax, most *Philogenia* species occur in the northern Andes, often very locally, with rarely more than two on the same stream (compare p. 233). *Archaeopodagrion* are found at forest streams there too, but generally at higher elevations. The species are similarly dark but smaller (under 1½ in/4 cm long) and more slender, with shorter legs and a distinctive stripy thorax. Males have diagnostic hornlike structures on the hind rim of the prothorax.

Genetics confirm that the two genera are close, but finding shared features to easily characterize the family's adults is difficult (p. 196). The males' claspers are structured similarly though, while the penis has two very long flagella, coiled like springs. The nymphs are similar too, with relatively long antennae and three swollen gills narrowing into a long filament.

LEFT | Male of the Golfo Dulce Flatwing (*Philogenia championi*) from Costa Rica.

BELOW | As most *Philogenia* species are similar and quite dull (some have dark wing tips), this male Blue Flatwing (*P. peacocki*) from Costa Rica is remarkably distinct.

FAMILY: RIMANELLIDAE
TEPUI DAMSELS

The tepuis are famed for the world's tallest waterfall (Venezuela's Angel Falls) and the isolated lifeforms surviving on their sandstone tabletops, which include South America's only Calopteryginae (p. 186). Although the Tepui Damsel (*Rimanella arcana*) occurs at low elevations in foothills too, east to French Guiana, it is restricted to the ancient Guiana Shield, especially around the tallest tepuis in Venezuela.

Its nymphs are among the strangest in Odonata, having two whiplike terminal appendages that are jointed and about as long as the body. These "tails" and their flattened bodies and legs make them look like stonefly nymphs (p. 204). They cling to root packs and wood in fast-flowing water, favoring rocks covered by riverweeds (Podostemaceae). Males perch nearby, usually with wings widely spread, but readily close them too.

Ecologically, this unique damselfly is remarkably like *Pentaphlebia* from Africa (see opposite). Even the red, robust, and long-winged adults are alike. The nymphs of both respire using gill tufts, rather than with their peculiar appendages, and were thus formerly considered related (p. 209). Genetics suggest a link to the tropical American families in the preceding profiles, though, so this exceptional species is placed in its own family.

LEFT | A male of the Tepui Damsel (*Rimanella arcana*) from Guyana.

DIVERSITY
1 species of rapid sections of rainforest streams and rivers in mountains and foothills in northern South America

ADULT HABIT
Medium-sized, rather robust damselflies, unique within their small range due to the largely red male abdomen and very densely veined wings that are clear and distinctly stalked (bases narrow and largely without cross-veins) with very long stigmas

TAXONOMY
Genus *Rimanella* (1 species)

FAMILY: PENTAPHLEBIIDAE

RELICS

Due to Africa's erratic climate (p. 54), few ancient odonate lineages survived there. While adult relics (*Pentaphlebia*) are distinct because of their considerable size and narrow but reticulate wings, the stonefly-like nymphs are truly strange, having prominent dorsal spines and terminal appendages shaped like a short trident flanked by two sickles, each with an angled "handle" and crescentic "blade," and gill tufts at their base.

While long associated with *Amphipteryx* (p. 209) and especially *Rimanella* (see opposite), genetically these damsels appear closest to the two Madagascan families treated next. Aside from Amanipodagrionidae (p. 217), they are the only odonate family restricted to Africa, being limited to Lower Guinea, the continent's wettest area, which is centered on Cameroon and Gabon.

While they can be deep red (one species is wholly black), males are very inconspicuous and inactive, perching with closed wings in the shade near small rapids and cascades. With their flattened bodies and legs, the nymphs cling to the undersides of rocks. Known only from a single male collected in 1973 on Nigeria's Obudu Plateau, the Obudu Relic (*P. gamblesi*) is the holy grail of African dragonfly exploration.

DIVERSITY
3 species of fast-flowing sections in rainforest streams and rivers in Central Africa

ADULT HABIT
Large damselflies unique within range for their size and long, narrow, densely veined, yellow-stained wings with darkened, slightly bent tips, and very long stigmas

TAXONOMY
Genus *Pentaphlebia* (3 species)

ABOVE LEFT | A male of the Red Relic (*Pentaphlebia stahli*) from Nigeria.

ABOVE | A nymph of the Red Relic (*Pentaphlebia stahli*) from Nigeria.

FAMILY: PROTOLESTIDAE AND TATOCNEMIDIDAE

PROTOS AND ROCKSTARS

Madagascar's nickname sadly alludes to the Great Red Island's denuded soils, which even color its rivers and the surrounding seas red. Being first in line for the Indian Ocean's rains, large areas actually remained wet and green for millions of years, and so many of the unique lifeforms here not only evolved and survived due to their long isolation but also because amenable conditions persisted.

Like many peculiar damselflies, over 30 species of running waters in Madagascar's forest remnants were long placed in Megapodagrionidae (p. 194).

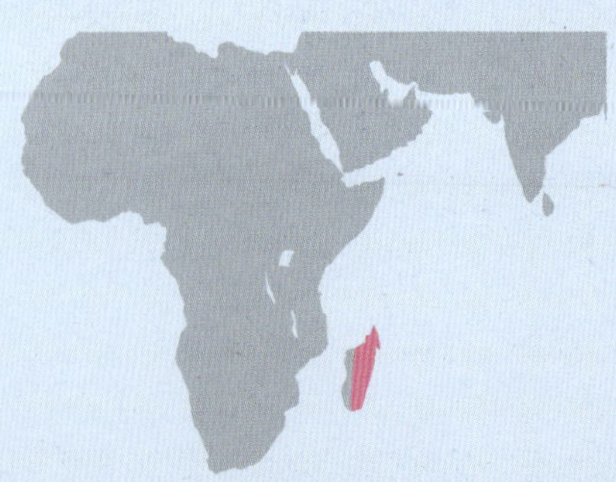

DIVERSITY
18 species of forested seeps and streams on Madagascar

ADULT HABIT
Medium-sized damselflies, distinct within range for perching by running forest waters with abdomen level and wings often (partly) spread and having either rather broad heads (Protolestidae), or 2 or 3 shallow "bites" out of the wingtip margin (Tatocnemididae)

TAXONOMY
Family Protolestidae with genus *Protolestes* (8 species) and Tatocnemididae with *Tatocnemis* (10)

LEFT | A male of the Common Rockstar (*Tatocnemis malgassica*).

RIGHT | The coloration of the male Rusty-tipped Proto (*Protolestes kerckhoffae*) matches that of several unrelated but co-occurring dragonflies (pp. 81, 99, and 108). Why appearances would converge like this on Madagascar is a mystery.

BELOW | Male of the White-nosed Proto (*Protolestes proselytus*).

Nesolestes was moved to Argiolestidae, its nearest relatives living in Gabon and the Seychelles (p. 198), but genomics found that the protos (*Protolestes*) and rockstars (*Tatocnemis*) have no close surviving relatives at all, so both have been recognized as families since 2022 (p. 216).

Although most damselflies perch with folded wings, and many former "megapods" spread them, these genera mix their posture up. Both keep the abdomen roughly horizontal, but *Tatocnemis* either closes or spreads the wings widely, depending on its state of agitation. Males sit motionlessly on rocks and plants, just above dark, stony-bottomed streams, and are very inconspicuous despite their distinctive red abdomens. *Protolestes* holds the wings closed to half-open; its species are found at a wider array of forest steams.

Most lifeforms on Madagascar occurs nowhere else, inspiring the Eighth Continent's other nickname. Not only is the same true for its odonates, but nowhere is the gap between our knowledge of the species and our fear for their fates greater: over 50 of the 172 recognized remain unreported since their first being named. Some may truly be gone, as two-fifths of suitable habitat disappeared between 1950 and 2000 alone, but no one has really looked hard enough either. Even more species probably remain unrecorded altogether, so over 200 may actually occur, with at least three-quarters unique to Madagascar.

FAMILY: PRISCAGRIONIDAE
XIANS

Odonates seem more varied in south China than anywhere else, particularly because (aside from numerous species) so many distinct evolutionary lineages are present. Probably, so many could survive here (and probably arise too) because habitats ranging from steamy pools in the lowlands to icy torrents on mountain tops could persist close-by and continuously across the vast and varied terrain for long periods.

The latest family to be added may be among the oldest, the proposed common name referring to the immortal beings of Taoist philosophy (compare p. 237). While genomics led to the recognition of quite a few new families in 2022 (p. 216), these xian were not even known until *Sinocnemis* was described in Platycnemididae in 2000 and *Priscagrion* in Megapodagrionidae a year later. The genetics, moreover, suggest their lineage branched off before all damselfly groups on the foregoing pages evolved. Only those treated next are even more distinct.

Spreadeagled on exposed surfaces (rocks, leaves), the robust *Sinocnemis* recalls *Mesopodagrion*, which is found alongside it (p. 216). They are more colorful, however, with bright blue (rather than pruinose) abdomen tips. Colored similarly but often with dark wingtips, *Priscagrion* males are larger and more slender, more like argiolestids (p. 196), perching in the shade by rocky forest streams with open wings and level body. Like the males' distinctly shaped penis and claspers, discovery of the nymphs will undoubtedly substantiate the link between these remarkable genera.

LEFT | A male Stout Xian (*Sinocnemis yangbingi*) from China.

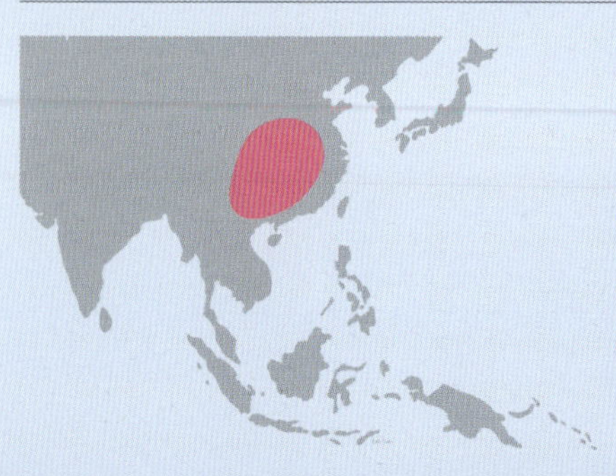

DIVERSITY
5 species of (mountain) forest streams in southern China and northern Vietnam

ADULT HABIT
Rather large damselflies with quite densely veined wings held open at rest, best separated from similar damsels within their small range by being quite robust and dark with bright markings; the last three abdomen segments are rather broad and notably blue

TAXONOMY
Genera *Priscagrion* (2 species) and *Sinocnemis* (3)

FAMILY: PLATYSTICTIDAE
SHADOWDAMSELS

As their name suggests, the mostly small and slender shadowdamsels typically lurk in the forest gloom. They often sit very close to the tiniest of flowing waters, hard to spot against the dark backdrop of rocks or leaf-litter, or deep in dense vegetation. Only when disturbed, hovering briefly before returning to their perch, might the slight touches of white, yellow, and blue on their black or brown bodies stand out.

After lestids and their relatives (pp. 238–249), platystictids form Zygoptera's most distinct lineage, diverging before all damsels treated on the previous pages did. Although diagnosed by details of their veins and minute ridges on the back of the head, the adults' frail habit is unique too: they have "pinched" wingtips, deep stigmas (*Platysticta* is Greek for "broad-spot"), and long, thin abdomens that are often held curved, as if weighed down (or lifted up) by their slightly expanded tip.

The equally distinctive nymphs look a bit like termites, with notably large and ovoid heads and masks. The three terminal appendages are somewhat swollen with long filaments at their ends. They live beneath rocks in rapid streams or spray zones, but probably among soggy detritus too.

Such habitats dry out easily due to climate change or loss of tree cover, so these damsels only survive in stably wet environments. The need to disperse and find new breeding sites is limited there, so populations readily become isolated and give rise to new species (p. 9). Many are limited to single islands or mountain ranges, therefore. Unknown species are indeed encountered constantly, with well over 300 likely to exist.

BELOW | Male of the Serendib Forestdamsel (*Platysticta serendibica*) from Sri Lanka (see next page).

DIVERSITY
284 species of (deeply) shaded running waters (especially forest streamlets, seeps, and waterfall spray) in the American tropics and from Sri Lanka to the Solomon Islands

ADULT HABIT
Small to fairly large damselflies appearing distinctly dark and delicate with a flat face, small thorax, narrow wings (with often pointy and slightly back-curved tips bearing clearly converging veins and rhomboidal stigmas), and a thin abdomen that is often long and gently arched or sharply bent up or down at its end

TAXONOMY
4 well-defined subfamilies are treated on the following pages

ASIAN SHADOWDAMSELS

Although worldwide between one and four in every ten species are at risk (p. 29), of the 59 odonate species and eight subspecies found only on Sri Lanka, 52 may face extinction. Nowhere are odonates in greater peril. As habitat destruction is the foremost threat everywhere, and forest stream species are most in danger, it is unsurprising that half of Sri Lanka's endangered taxa are platystictids (see previous page). While placed in two genera, the larger forestdamsels (*Platysticta*), often with dark wingtips (photo p. 233), probably evolved from among the small *Ceylonosticta* species.

Today, only the third that covers the island's windward side and its highlands receives enough rain to support its unique shadowdamsels. This area, the size of a small country such as Slovenia, is also home to most of Sri Lanka's 22 million people, however. These damsels' lineage separated before all other extant platystictids diversified, moreover, except for the even older Sinostictinae (p. 237). This separate radiation, classified as its own subfamily Platystictinae, and that in tropical America (p. 236), suggests that once platystictids lived in Africa and Madagascar too. These were then lost to the changeable climate (p. 54), long before humans arrived on the scene.

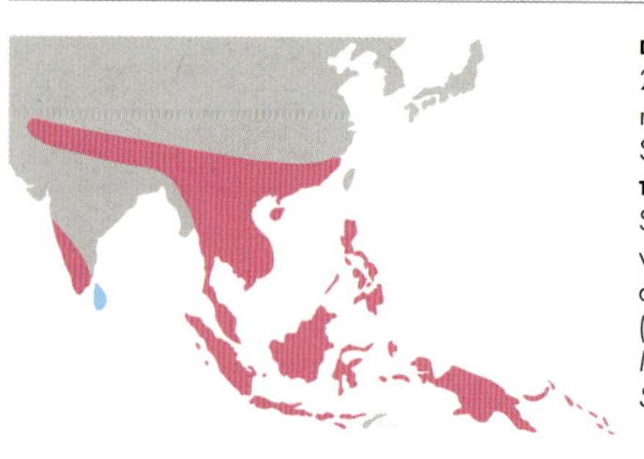

DIVERSITY
234 species of mostly smaller, shady running waters from Sri Lanka to the Solomon Islands

TAXONOMY
Subfamily Platystictinae (blue on map) with genera *Ceylonosticta* (22 species) and *Platysticta* (4); and Protostictinae (pink) with *Drepanosticta* (127), *Indosticta* (1), *Protosticta* (58), *Sulcosticta* (5), and *Telosticta* (17)

ABOVE LEFT | Female of the Drooping Shadowdamsel (*Ceylonosticta lankanensis*) from Sri Lanka.

OPPOSITE BOTTOM | Male of the Khao Soi Dao Shadowdamsel (*Protosticta khaosoidaoensis*) from Thailand.

LEFT | A male of Emtra's Shadowdamsel (*Drepanosticta emtrai*) from Vietnam.

The most isolated member of this family of rainforest hermits is *Drepanosticta palauensis* from Babeldaob. Barely 25 miles (40 km) long, its island in the Micronesian nation of Palau is at least 530 miles (850 km) from the next suitable habitat. Unfortunately, ascertaining whether this castaway damselfly originated in the equidistant southern Philippines, northern Moluccas, or western New Guinea is complicated by the taxonomic state of its subfamily, Protostictinae.

Currently, the more than 200 species named are placed in just five genera. Perhaps they evolved so rapidly that it is effectively impossible to find clear characters that divide them into distinct groups. Many species, for example, can only be distinguished by the males' complex claspers—often with great difficulty.

Venational distinctions used to define the genera so far are just too small and variable: some individuals match *Protosticta* in one wing, but *Drepanosticta* in another! Experts are gradually redefining the groups, therefore: distinctive species from Luzon in the Philippines were split off as *Sulcosticta*, some from Borneo and Palawan as *Telosticta*, while India's single *Platysticta* species (see above) was moved to *Indosticta*.

Each new genus is characterized by subtle combinations of its species' markings, male claspers and penis, and even the structure of the head and thorax. The prothorax in *Telosticta* bears lobes that recall the drooping ears of a spaniel, for example. *Drepanosticta* may be limited to mainland Asia and *Protosticta* to Sulawesi, from where their type species originate. The numerous species on New Guinea and many other islands, including Babeldaob, may well represent further unnamed genera.

PLATYSTICTIDAE—PALAEMNEMATINAE

AMERICAN SHADOWDAMSELS

American platystictids seem to have broader preferences than their Old World counterparts, with larger and more robust adults, and comparatively broader wings. Central America's widespread Nathalia Shadowdamsel (*Palaemnema nathalia*) is common on small rocky rivers outside of forests, for example, while the Desert Shadowdamsel (*P. domina*) extends from Nicaragua to the drylands of Arizona.

Nonetheless, most species are partial to rainforest streamlets, seeps, and spray. All, moreover, depend on permanent and deeply shaded water, and seek the cover of dense thickets and tangles where mountain streams descend into the desert, or tree buttresses in open pasture. Many of the American species, indeed, also occur only locally, although up to nine have been found together in Costa Rica!

Solitary males may hang near water to defend a small territory, but some species form leks, gathering among shrubs over a stream to await females. They spread the wings widely and then snap them shut. This may fend off rivals but also attract mates, and is presumably emphasized by the dark wingtips found in many species. Whether other platystictids, such as the similar *Platysticta* from Sri Lanka (p. 233), have comparable habits remains to be discovered.

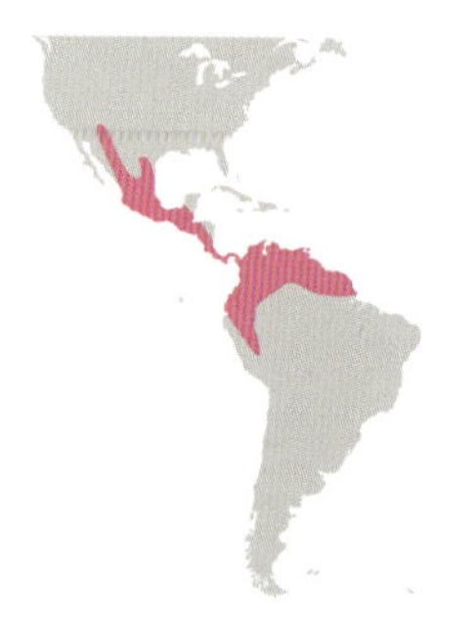

DIVERSITY
43 species of shaded running waters in Middle America and northern South America

TAXONOMY
Genus *Palaemnema* (43 species)

ABOVE TOP | Male of the Desert Shadowdamsel (*Palaemnema domina*) from Arizona, USA.

ABOVE LOWER | Pair of Atoyac Shadowdamsels (*Palaemnema paulitoyaca*) from northeastern Mexico.

PLATYSTICTIDAE—SINOSTICTINAE

PIXIUS

The immortal xian (p. 232) are protected by Pixiu, mythical beings with a dragon's head, lion's body, and bird's wings. Being more robust than the other platystictids, *Sinosticta* species seem like hybrid creatures too, with their slightly broader wings and clearly shorter abdomens that are held level and sometimes even slightly raised at rest, but typically with the tip curved down. Also being quite colorful and perching with closed wings on top of leaves rather than hanging off their tips, they are easily mistaken for a member of another family, therefore, such as Rhipidolestidae (p. 214).

Posture and color do not define the subfamily, however, which is evolutionarily the most distinct in this already very distinct damselfly family (p. 233). *Yunnanosticta* was described from the Chinese province of Yunnan, close to the Myanmar border, only in 2017. Both species are tiny, slim, and almost completely dark brown but for the pale blue and strongly downcurved abdomen tip. Venation and claspers, nonetheless, match *Sinosticta*. A new *Yunnanosticta*, found in India's extreme northeast only in 2022, looks more like *Sinosticta* again, showing just how much remains to be discovered and understood about the shadowdamsels.

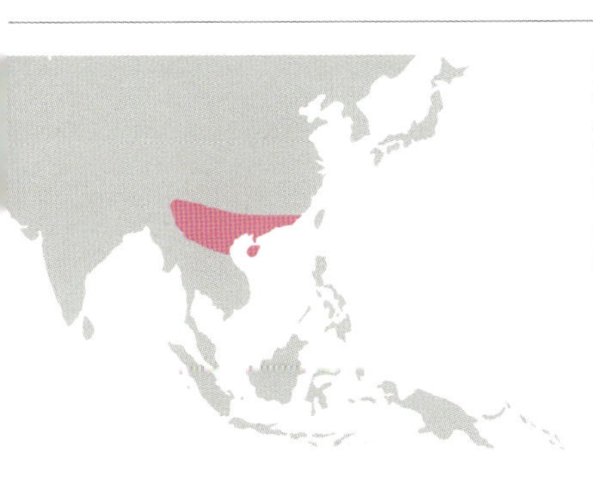

DIVERSITY
7 species of streamlets, trickles, and seepages in southern China's upland forests, extending just into Vietnam and Northeast India

TAXONOMY
Genera *Sinosticta* (4 species) and *Yunnanosticta* (3)

ABOVE TOP | The male Yellow-spotted Pixiu (*Sinosticta ogatai*) from China is comparatively similar to some shadowdamsels.

ABOVE LOWER | Some *Sinosticta* species, such as this male Hainan Pixiu (*S. hainanense*), are among the most stout and colorful platystictids.

FAMILY: LESTIDAE

SPREADWINGS

Besides the dominant Coenagrionidae (p. 132), this is the main group of damselflies in standing waters. With the families treated next, lestids set off on a separate evolutionary path before all other extant damselfly lineages diverged. The rise of pond-dwelling damselflies mirrors that of the dragonflies, therefore, in which only the distantly related Aeshnidae rival the overwhelming success of Libellulidae (p. 20). Lestids took a comparably distinct approach to that of their dragonfly counterparts (p. 112), focusing on the rich but often sporadic pickings provided by relatively marginal habitats, such as temporary waters. Derived from the Greek for "robber," the name *Lestes* is therefore quite appropriate!

LEFT | A pair of Common Spreadwings (*Lestes sponsa*) in tandem in the UK.

TOP RIGHT | Nymph of Europe's Migrant Spreadwing (*Lestes barbarus*) showing its long mask and massive gills.

RIGHT | Northern Australia's Slender Reedling (*Indolestes tenuissimus*) male is among a third of lestids that perch with closed wings.

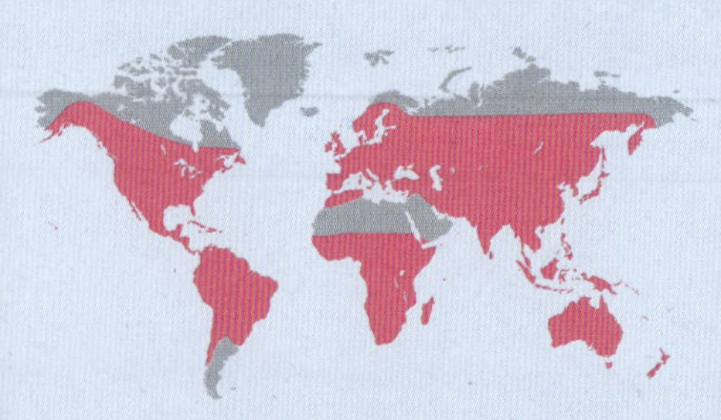

DIVERSITY
147 species of a wide variety of standing and running waters worldwide, notably temporary ponds

ADULT HABIT
Medium-sized to large damselflies, unique in many parts of the world and particularly at standing waters for combining larger size, wider heads, irregular mesh of wing veins, long rectangular stigmas, pincerlike male claspers, often extensive metallic and/or pruinose markings, and spread wings and hanging abdomens at rest

TAXONOMY
No subdivision of this large family is possible as yet (see p. 242), so 3 groups of similar genera are treated separately; forms the superfamily Lestoidea with Synlestidae (p. 244), Perilestidae (p. 248), and Hemiphlebiidae (p. 249)

Over two-thirds of the species perch with the wings spread, which facilitates rapid takeoff. This can be seen as the damselfly equivalent of the sustained flight of some dragonflies, allowing them to react quickly, providing an advantage especially when densities of prey, rivals, and mates are high. Life at temporary ponds can be particularly frenetic, and few males accidently grab females of other species as often as *Lestes* do!

The long-legged nymphs are built to grow big quickly. As in "cavilabiate" dragonflies (p. 86), the mask's palps are expanded, recalling hands with spread fingers. In many *Lestes*, the mask itself is exceptionally long and narrow, moreover, providing extra reach to gobble up the abundant prey in warm shallows. The three leaflike gills are notably large, probably to absorb enough oxygen from stagnant water, or provide a quick escape.

To profit from ephemeral opportunities, adverse periods in-between must be overcome. Spending months hanging around depressions that eventually fill up again with rain, lestids may have some of the longest adult lifespans; bright colors develop only once breeding starts (photo p. 11). The hardiest "spreadwings" actually shut their wings consistently (pp. 242–3), probably to be less visible to predators (p. 194).

Where winters are cold, the eggs are often the longest-lived life-stage. Laid late in summer, they only hatch once the water heats up again in spring, with adults emerging a few months later, often in great numbers to feed on the abundant flies and midges. Most lestids oviposit in sturdy vegetation, often well above water, where the eggs are well-protected.

LESTIDAE—*LESTES* AND SIMILAR GENERA

TYPICAL SPREADWINGS

To puncture the plants that must protect their eggs (see previous page), many lestids have strong ovipositors with two serrated plates cutting a small opening to receive each one. Indeed, of the rather few odonates that penetrate living bark, many belong to this family. Western Eurasia's willow spreadwings (*Chalcolestes;* **purple on map**) oviposit in woody plants overhanging garden ponds, for example, while the various *Lestes* species in the drying-out marshlands nearby use rushes and sedges.

Found dangling from the branches overhanging deeply-shaded pools, such as dark forest recesses flooded after rain, southeast Asia's giant spreadwings (*Orolestes;* **also purple**) are the most spectacular lestids. Almost 3 in (8 cm) long, the male's dark-metallic body is marked exquisitely with pale green and blue. Although clear-winged forms occur too, their wings are patterned with black blotches, sometimes fringed with white (compare pp. 182 and 213). The nymphs must feed voraciously to grow so big quickly, and have huge palps on very long-stalked masks (see previous page).

Despite appearing nearest *Orolestes* genetically, the *Chalcolestes* species remained in *Lestes* for almost two centuries. Today, over half of lestids still reside

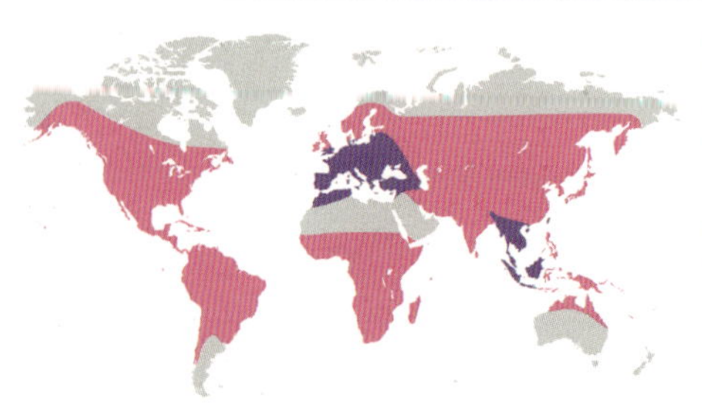

DIVERSITY
99 species of a wide variety of freshwaters (often but not exclusively temporary, also flowing) worldwide; absent only in polar regions

TAXONOMY
Genera *Archilestes* (9 species), *Chalcolestes* (2), *Lestes* (80), *Orolestes* (3), *Platylestes* (4), and *Sinhalestes* (1)

ABOVE LEFT | Male of the Western Willow Spreadwing (*Chalcolestes viridis*) in the UK.

there, although the 80 species are virtually cosmopolitan and incredibly diverse. Many adults are extensively metallic, pruinose, or both, for example, while others are largely dull-brown or blue and black (photo p. 11). And although temporary ponds are favored, their habitats vary widely in shading, vegetation, and longevity.

This hodgepodge may eventually be divided across many more genera, so the names used currently might prove quite uninformative. Ranging from the Rockies to Rio, stream spreadwings (*Archilestes*) may only have their great size in common, for example. And that might just be a shared adaptation to running waters, where they too oviposit in woody stems. With their square stigmas, tropical Asia's *Platylestes* seem undistinctive too. The endangered Sri Lanka Spreadwing (*Sinhalestes orientalis*), finally, was unrecorded from 1858 until its rediscovery in 2012 (p. 234). Being green-metallic and inhabiting rocky forest streams, it recalls *Archilestes*, and synlestids too (p. 244).

ABOVE | Male of the Eight-spotted Giant Spreadwing (*Orolestes octomaculatus*) in Thailand guarding an ovipositing female.

BELOW | A pair of California Spreadwings (*Archilestes californicus*) laying eggs into a twig.

RINGTAILS AND REEDLINGS

When cold or disturbed, even spread-winged damselflies quickly fold their wings, as do freshly emerged and roosting adults. Indeed, the lestids that may spend the longest and harshest times waiting to breed consistently shut their wings when perched, typically slumping them beside the abdomen (compare pp. 238–41).

While these "foldwings" were once grouped as the subfamily Sympecmatinae, genetics suggest that only *Sympecma* (see opposite) is near *Lestes*, and not the remaining species. Most are classified as reedlings (*Indolestes*, photo p. 239), but the ringtails (*Austrolestes*) from Australia to New Zealand may well have evolved from among them. While they occur from rainforests to deserts, and from sea-level to high plateaus, many species are abundant only locally, suggesting they are the edgiest in a family used to living on the edge.

Australia's Inland Ringtail (*A. aridus*) lives only in the baking outback, for example, while the Cave Reedling (*I. obiri*) breeds in the Top End's seasonal rock pools, adults surviving under overhangs and in shallow caves. Most other *Indolestes* are poorly known but some Asian species favor mountain ponds and hibernate as adults like *Sympecma*. Some New Guinean ones may have peculiar reproductive behavior, as males possess weirdly expanded (sometimes hairy) hind wing bases.

Aside from posture, these damselflies also converged with coenagrionids in color. While immature males are often brown, orange, pink, or whitish, all species may ultimately become blue with dark markings. Perching above ponds, they are surprisingly like bluets (p. 148). While most odonates change color gradually and unidirectionally with age, an *Indolestes* species was proven to turn blue within minutes when warming up, reversing to brown within hours once cooling off again.

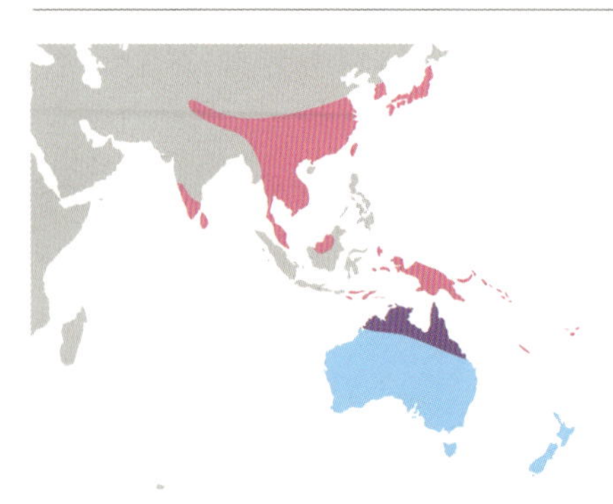

DIVERSITY
45 species of standing waters almost anywhere from Japan and Sri Lanka to New Zealand, often in more extreme environments such as deserts and high plateaus

TAXONOMY
Genera *Austrolestes* (10 species; **blue on map**) and *Indolestes* (35; **pink**)

ABOVE LEFT | A mating pair of Metallic Ringtails (*Austrolestes cingulatus*) from Australia.

LESTIDAE — *SYMPECMA*

WINTER DAMSELS

Compared with many insects, odonates like it hot (compare p. 115). Waters that are too cold for them will harbor multiple stonefly and caddisfly species, for example. The species that do occur in temperate and subpolar climes largely survive the winter as egg or nymph, when they are comparatively sheltered.

By odonate standards, lestids are extremophiles (see previous pages). Like the co-occurring *Lestes* species, winter damsel (*Sympecma*) adults emerge in the course of summer. Rather than maturing rapidly to mate and lay eggs, they spend the following months feeding. Once it gets colder, these damselflies settle in the vegetation, pressed against their perch with the wings held alongside the abdomen. While they also creep under stones or bark, they often remain remarkably exposed, relying on their cryptic coloration to avoid predation.

Although these damselflies may be covered completely with snow, they reactivate quite quickly when it is sunny, and sometimes fly about on pleasant winter days. Reproduction only starts in earnest in early spring, however. Unhindered by other odonates and birds feeding their hungry young, the eggs are laid in the soft and often floating remnants of last year's vegetation.

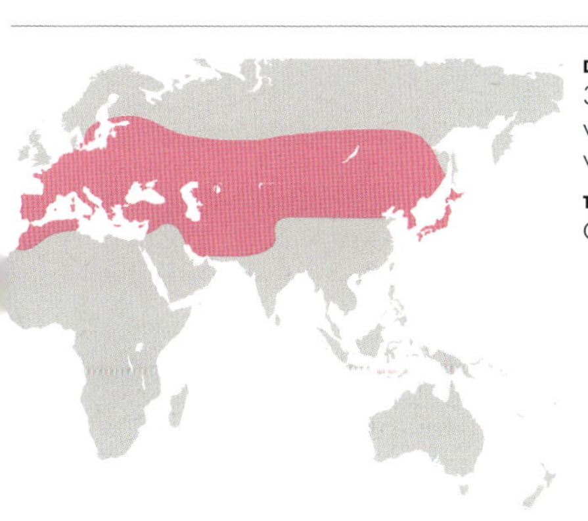

DIVERSITY
3 species of various open standing waters (mostly shallower and well-vegetated) from Morocco to Japan

TAXONOMY
Genus *Sympecma* (3 species)

ABOVE TOP | Hibernating female Siberian Winter Damsel (*Sympecma paedisca*).

ABOVE LOWER | A young Common Winter Damsel (*Sympecma fusca*). Unlike other "foldwings" (see opposite), adults at most attain some blue in the eyes.

FAMILY: SYNLESTIDAE
MALACHITES

LEFT | As its common name suggests, eastern Australia's Southern Whitetip (*Episynlestes albicaudus*) is notable for its bright abdomen tip.

TOP RIGHT | Variable Shujings (*Sinolestes editus*) in China about to mate.

LOWER RIGHT | Flitting with their white-banded wings in forest shade or over blackwater rivers, *Chlorolestes* species such as this Mountain Malachite (*C. fasciatus*) from South Africa have been mistaken for elves!

Each globally distributed odonate family that dominates stagnant waters today has more localized and inconspicuous relatives in streams (p. 20): for example, Libellulidae and Synthemistidae, Aeshnidae and Austropetaliidae, and Coenagrionidae and Platycnemididae. The spotty distribution of Synlestidae suggests that they once were also almost as widespread as Lestidae.

Living near Brisbane over 200 mya, the oldest odonate nymphs known were remarkably like the synlestids found there today. As late Triassic adults cannot be assigned to any extant suborder with certainty, it is highly unlikely that these nymphs are actually related. The likeness does suggest, however, that odonates have lived similar lives here for an exceptionally long time. Being very short and round, the leaflike terminal gills, especially, are uncannily alike.

Compared with lestids, synlestids need only small gills to respire, as cool moving water is highly aerated. Large gills are also impractical in the current. Nymphs may never have needed a tailfin

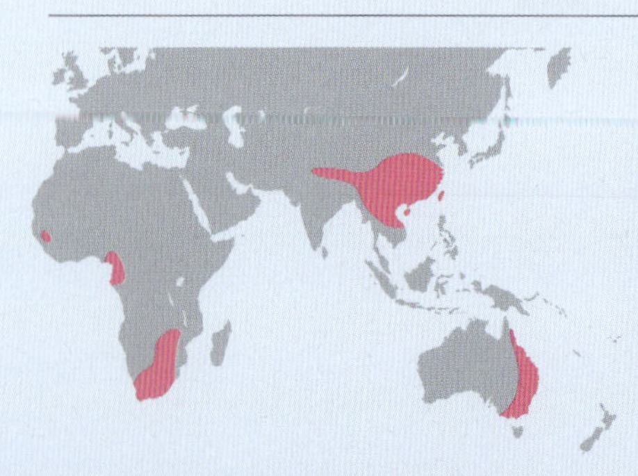

DIVERSITY
31 species of calmer sections of streams and sometimes rivers, especially in cooler parts (mountains, forests) of eastern Australia, southern and eastern Asia, southern and western Africa, and Hispaniola

ADULT HABIT
Rather large damselflies recalling lestids; within their range most damsels with (usually) green-metallic bodies perching at cool streams with wings spread and long abdomen hanging down will be synlestids

TAXONOMY
The geographically widely disjunct genera probably represent 3 distinct subfamilies, with *Megalestes* (12 species) and *Sinolestes* (1) in Asia; *Chorismagrion* (1 species), *Episynlestes* (3), and *Synlestes* (3) in Australia; and *Chlorolestes* (7 species),

(nor the sense!) to evade fish, as they often sit on rocks, out in the open. The mask is smaller too, as less reach is needed among rocks and detritus than in the lestids' three-dimensional world of aquatic vegetation (compare p. 239).

The adults of the two families look surprisingly similar, by contrast, typically resting with the wings spread widely and abdomen hanging down. Aside from differing in the details of the wing veins and favoring open standing waters, most lestids within the synlestids' range are not so extensively green-metallic, however.

East Asia may harbor more ancient and isolated odonate diversity than anywhere else (p. 232) and, with most species concentrated there, this relict family forms no exception. As their names indicate, the Asian malachites (*Megalestes*) are rather large synlestids (usually 2½–3⅛ in/6.5–8 cm long) with green-metallic bodies, often marked boldly with yellow.

With comparatively long masks, as well as long and thin legs, the spindly nymphs are typically found at mountain streams, which are often rocky and shaded by forest. The rather sluggish males hang conspicuously on branches above calmer sections (such as springs and headwater pools), quickly resettling nearby after being disturbed.

The equally big Variable Shujing (*Sinolestes editus*) ranges from south China to Taiwan, Hainan, and Vietnam. Males can be clear-winged like females and *Megalestes*, but some have variable

Ecchlorolestes (2), *Nubiolestes* (1), and *Phylolestes* (1) in Africa and Hispaniola brown bands below the stigmas, while others have black wings with milky tips and only their bases clear, recalling the distantly related Selys's Giant Spreadwing (photo p. 21).

Occupying another hotspot for archaic odonates, Australia's Pretty Relict (*Chorismagrion risi*) and needles (*Synlestes*) are rather unshowy, the males perching quietly by stream pools. The whitetips (*Episynlestes*) stand out due to the males' bright claspers, however. These are not only very pale, but exceptionally long, with a high knob at their base adorned with a peculiar tuft of hair in two species.

Southern Africa's nine synlestids are remarkably varied considering their limited range. Of five species confined to the Cape, for example, the Queen Malachite (*Ecchlorolestes nylephtha*) hangs unnoticed near dark forest trickles on the moist south coast, like a green-metallic thread. On the dry west coast, the gray-marbled Rock Malachite (*E. peringueyi*) is hidden too, pressed against rugged boulders in open streams.

By contrast, five of the seven *Chlorolestes* species, which extend from the scrubby fynbos of the Cape to Mozambique's upland forests and grasslands, develop conspicuous dark wing bands, and often bands of white pruinosity too. Some males remain clear-winged, however, which may allow them to intercept the potential mates of their territorial banded brothers (compare pp. 182 and 213).

Largely restricted to Gabon and Cameroon, the Rainforest Malachite (*Nubiolestes diotima*)

hangs motionlessly by shady mountain streams, often with wings only half open, just its very long abdomen's widened white-pruinose tip standing out in the gloom. While long seen as the Old World's sole perilestid (next profile), genetic evidence suggests it is close to *Chlorolestes*, as well as to the single extant synlestid in the New World.

Despite its prolonged isolation, the Hispaniolan Malachite (*Phylolestes ethelae*) looks much like its relatives, and similarly breeds in pools in mountain forest streams, where males dangle off twigs and fronds nearby. Reproductive behavior may be concentrated at dusk. Breeding in darkness would be unique among odonates, although many (localized) groups have yet to be studied.

Fossils from Canada and Patagonia suggest that synlestids were once much more widespread in the Americas. A second but still unnamed *Nubiolestes* species, encountered by the author in Guinea in 2019, moreover, narrowed the 5,000 mile (8,500 km) gap to Africa by 1,500 miles (2,500 km)! Such localized and ancient species are usually threatened, *P. ethelae* being ranked as Endangered on the IUCN Red List, for example.

LEFT | Female of the Hispaniolan Malachite (*Phylolestes ethelae*) that, unlike its Old World relatives, always closes the wings at rest.

ABOVE | This male Rainforest Malachite (*Nubiolestes diotima*) from Nigeria is a comparatively colorful representative of the family.

RIGHT | A male Rock Malachite (*Ecchlorolestes peringueyi*) perched flat on a rock surface.

FAMILY: PERILESTIDAE

TWIGTAILS

This group is the counterpart of Synlestidae (previous profile) on the American mainland. Similarly found among leaf-litter in quiet stream pools (such as below small waterfalls), the nymphs are almost identical. Adults look quite different, though, their abdomens being even longer and often well over twice as long as the wings, making the latter appear especially short. Each segment bears one or two pale rings, while the thorax is contrastingly striped, so perilestids are patterned dark and light rather than mostly looking bronzy green.

Males are encountered sparingly and are easily overlooked as they hang from branches or leaf tips up to 3 ft (1 m) above the water, usually with half-open wings. The vertical abdomen's tip is often curved upward, making the exceptionally lengthy damsels look like twigs or vines.

While it has been proposed to merge the two families, the latest evidence suggests that they have a common ancestor but separate evolutionary histories, so the present classification may as well be preserved. The genera *Perilestes* and *Perissolestes* are so alike that all species may be grouped under *Perilestes* following further study, however.

DIVERSITY
21 species of small rainforest streams in the American tropics

ADULT HABIT
Fairly large damselflies recognized within range by exceptionally long pale-ringed abdomen that hangs down beneath half-open wings at rest

TAXONOMY
Genera *Perilestes* (9 species) and *Perissolestes* (12)

ABOVE LEFT | Female of the Horned Twigtail (*Perissolestes magdalenae*).

FAMILY: HEMIPHLEBIIDAE
GREENLINGS

Believed to have set off on its evolutionary path before all other families of the most distinct ineage of damselflies evolved, the Ancient Greenling (*Hemiphlebia mirabilis*) is often seen as the most primitive living odonate (p. 238). Its wing venation is uniquely simple, with similar fossils known from the Cretaceous of Eurasia and South and possibly North America. Perhaps, therefore, this tiny damselfly is all that remains of a group occurring globally in the age of dinosaurs. While long thought to be very rare, and even threatened with extinction, more populations were discovered recently. Adults live barely a week but are very abundant, with over a million estimated to emerge each year at one site!

The lower pair of claspers on the male's abdomen tip is unique, being leaflike and brightly blue-white. Even stranger is that females carry such appendages too! When perched, both flash them by frequently flicking the abdomen up in a curve. Males continue flashing even as they attempt to mate. Mating was rarely observed and egg-laying never seen, even during prolonged studies. Might these happen at night (compare p. 247)? More mysteries about this supposed living fossil, one of many in an apparently ancient insect order, therefore remain.

DIVERSITY
1 species of marshes with dense reeds and sedges (often dried out in summer) on Tasmania and adjacent Australia

ADULT HABIT
Tiny (about 1 in/2.5 cm long) green-metallic damselfly with unique characters such as a pair of leaflike blue-white appendages on both sexes' abdomen tip

TAXONOMY
Genus *Hemiphlebia* (1 species)

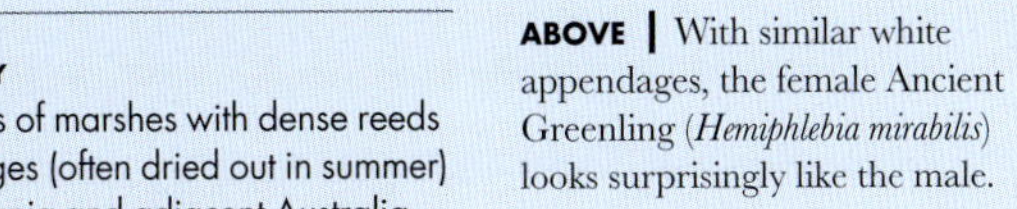

ABOVE | With similar white appendages, the female Ancient Greenling (*Hemiphlebia mirabilis*) looks surprisingly like the male.

GLOSSARY

abdomen: tail-like section of body posterior to thorax

Afrotropics: the tropical realm that includes Africa south of the Sahara and nearby islands

Anisoptera: the suborder in which typical dragonflies (thus all odonates except damselflies and damseldragons) are placed

appendages: structures at the abdomen tip with various functions, including as claspers in all male odonates and as gills in many damselfly nymphs

auricle: ear-like protrusion on the sides of abdomen base in some male dragonflies

cercus (pl. cerci): paired upper appendage on abdomen tip

claspers: modified appendages at abdomen tip used by males to grasp their mates, see p. 17

crepuscular: the habit of being most active in twilight

cross-veins: short veins that are perpendicular to more continuous longitudinal veins of wing

detritus: organic matter such as leaf-litter

endemic: being confined to a limited part of the world, such as an island

epiproct: lower appendage on abdomen tip, notably in claspers of male dragonflies

family: most commonly applied taxonomic grouping of closely related genera

femur (pl. femora): innermost of two longer sections of the legs, i.e. the "thighs"

genus (pl. genera): taxonomic grouping of closely related species, denoted by genus name

genus name: first part of species name, such as *Pantala* in *Pantala flavescens*

Gondwana(land): southern prehistoric supercontinent that formed after breakup of Pangea, see p. 111

holotype: specimen from which a species is originally described

mandible: gnawing mouthparts besides the "upper and lower lips" (i.e. labrum and labium)

mask: labium (lower lip) of odonate nymph, which is modified into a prehensile grasping arm, see p. 8

melanin: a common pigment, which is black in its densest form

mimic, mimicry: to look deceptively like another taxon, such as a wasp or unpalatable butterfly

Neotropics: the (mostly) tropical realm that includes Central and South America and nearby islands

niche: the ecological space inhabited by individuals of a species

node: a break or kink roughly halfway along the wing's leading edge

nymph: the developmental stage between egg and adult, often called larva and sometimes naiad too

oviposition: egg-laying behavior (either with or without an ovipositor)

ovipositor: apparatus below female's abdomen tip to cut into a substrate and place eggs there

Paleotropical: pertaining to the Old World (i.e. Africa, Eurasia, Australasia) tropics

palps: pair of finger- or hand-like graspers at tip of nymph's mask, see p. 8

paraproct: paired lower appendage on abdomen tip, notably in claspers of male damselflies

pigmentation: color created by pigments but not by pruinosity, see p. 13

postocular spot: pale marking beside eyes at back of damselfly's head

prothorax: small front section of the thorax (carrying the front legs) that connects to the head

pruinose, pruinosity: microscopic scales of wax that develop on the body, which reflect light in shades of white, gray, and blue

pterostigma: a thickened and often distinctly colored spot on leading edge of wing close to its tip

radiation: a notable evolutionary diversification of species, usually within a distinct space or time

relict: isolated representative of an ancient evolutionary lineage

seep, seepage: a place where water oozes from the ground

seta (pl. setae): hairlike bristles

species name: unique combination of a genus name (e.g. *Pantala*) and specific epipithet (e.g. *flavescens*) to name a species, such as *Pantala flavescens*

specific epithet: second part of species name, such as *flavescens* in *Pantala flavescens*

stigma: abbreviation of pterostigma, a thickened and often distinctly colored spot on leading edge of wing close to its tip

subfamily: subgroup of a taxonomic family, e.g. Aeshninae and Telephlebiinae are subfamilies of Aeshnidae

suborder: subgroup of a taxonomic order, e.g. Zygoptera (damselflies) is a suborder of the order Odonata

superfamily: grouping of related taxonomic families, e.g. the superfamily Aeshnoidea combines the families Aeshnidae and Austropetaliidae

synonym: in taxonomy a name that is predated by another name for the same species or taxon

tandem: position in which a male grasps a female with his appendages but they are not connected by their genitalia

taxon (pl. taxa): any named group of related organisms, such as a species or family

thorax: section of body that carries the wings and legs

tibia (pl. tibiae): outermost of two longer sections of the legs, i.e. the "shins"

tribe: subgroup of a taxonomic subfamily, e.g. Hemigomphini is a tribe of the subfamily Lindeniinae within the family Gomphidae

type species: the species selected to characterize a genus

venation: the network of veins in the wings

Zygoptera: the suborder in which all damselflies are placed

FURTHER READING AND USEFUL RESOURCES

ODONATE BIOLOGY

aulson, D.R. *Dragonflies and Damselflies: A Natural History.* y Press, 2019.

Corbet, P.S. *Dragonflies. Behaviour and Ecology of Odonata.* Harley Books, 1999.

Cordoba-Aguilar, A., C. Beatty & J. Bried (editors). *Dragonflies and Damselflies: Model Organisms for Ecological and Evolutionary Research.* Oxford University Press, 2022.

amways, M.J. *Conservation of Dragonflies: Sentinels of Freshwater Conservation.* CABI Publishing, 2024.

AMERICAS

Bota-Sierra, C.A., J. Sandoval-H, D. Ayala-Sánchez & R. Novelo-Gutiérrez. *Dragonflies of the Colombian Cordillera Occidental, look from Tatamá.* Published independently, 2019.

Lam, E. *Dragonflies of North America.* Princeton Field Guides, 2024.

Lompier, T. *A Guide to the Dragonflies and Damselflies of the Serra dos Orgaos, South-Eastern Brazil.* REGUA, 2015.

Paulson, D.R. *Dragonflies and Damselflies of the West.* Princeton Field Guides, 2009.

Paulson, D.R. *Dragonflies and Damselflies of the East.* Princeton Field Guides, 2011.

Paulson, D.R. & W.A. Haber. *Dragonflies and Damselflies of Costa Rica: A Field Guide.* Cornell University Press, 2021.

AFRICA AND EUROPE

Dijkstra, K.-D.B. & V. Clausnitzer. *The Dragonflies and Damselflies of Eastern Africa: handbook for all Odonata from Sudan to Zimbabwe.* Studies in Afrotropical Zoology, 2014.

Dijkstra, K.-D.B. & C. Cohen. *Dragonflies and Damselflies of Madagascar and the Western Indian Ocean Islands.* Association Vahatra, 2021.

Dijkstra, K.-D.B., A. Schröter & R Lewington. *Field Guide to the Dragonflies of Britain and Europe. Second Edition.* Bloomsbury Wildlife Guides, 2020.

Tarboton, W. & M. Tarboton. *A Guide to the Dragonflies & Damselflies of South Africa.* Struik Nature, 2015.

ASIA AND AUSTRALASIA

Kalkman, V.J. & A.G. Orr. *Field Guide to the Damselflies of New Guinea.* Brachytron Supplement, 2013.

Orr, A.G. *A Guide to Dragonflies of Borneo: Their Identification and Biology.* Natural History Publications (Borneo), 2003.

Orr, A.G. *A Pocket Guide: Dragonflies of Peninsular Malaysia and Singapore.* Natural History Publications (Borneo), 2005.

Orr, A.G. & V.J. Kalkman. *Field Guide to the Dragonflies of New Guinea.* Brachytron Supplement, 2015.

Ozono, A., I. Kawashima & R. Futahashi. *Dragonflies of Japan.* Bunichi-Sogo Syuppan, 2012.

Reels, G. & H. Zhang. *A Field Guide to the Dragonflies of Hainan.* China Forestry Publishing, 2015.

Singh, D. *Field Guide to the Dragonflies & Damselflies of Northwest India.* Bishen Singh Mahendra Pal Singh, 2022.

Tang, H.B., L.K. Wang & M. Hämäläinen. *A Photographic Guide to the Dragonflies of Singapore.* Raffles Museum of Biodiversity Research, 2010.

Theischinger, G., J. Hawking & A.G. Orr. *The Complete Field Guide to Dragonflies of Australia. Second Edition.* CSIRO Publishing, 2021.

Zhang, H. *Dragonflies and Damselflies of China. Vol. 1-2.* Chongqing University Press, 2019.

USEFUL WEB SITES

Worldwide Dragonfly Association and *International Journal of Odonatology*
www.worlddragonfly.org

Odonatologica and *Notulae Odonatologicae*
www.odonatologica.com

World Odonata List
www.odonatacentral.org/app/#/wol

The author's personal site
www.kddijkstra.nl

INDEX TO FAMILIES AND GENERA

PICTURE CREDITS

The publisher would like to thank the following for permission to reproduce copyright material: T = Top; B = Bottom; L = Left; R = Right; C = Center

Adolfo Cordero Rivera 183, 224L, 246.
Alamy/ Agami Photo Agency 173, Biosphoto 60, Blickwinkel 61, 171T, 206B, Bryan Reynolds 152B, Danita Delimont 17, 116, Gerry Pearce 149T, 242, Gillian Pullinger 238, Hakan Soderholm 243T, imagebroker.com GmbH 38T, John Corso 63B, Luis Louro 65B, 80B, Manjeet & Yograj Jadeja 137T, Minden Pictures 132, Rachel Kolokoff Hopper 51B, Steffan Rotter 74.
Alan Schmierer 66B; Alandmanson 172T; Allan Brandon 168; Alpsdake 87T; André Günther 186B, 189T, 189B; Andrew Allen 84; Andrew Lai 237T; Andrew Skinner 55T; Antoine van der Heijden 102T; Axel Hochkirch 165; Benoît Guillon, www.meslibellules.fr/blog/ 83T, 154T, 164T; Bert Harris 225T; © Callan Cohen/www.birdingafrica.com 199T, 231B; Cameron Eckert 51T, 67B, 159; Charles J. Sharp 3, 45B, 47B, 64, 70T, 71TL, 75B, 112, 148, 221T; Cheongweei Gan 190; Chris Chafer 129; Chris Kex 76; Christophe Brochard 9TC. 9TR, 9 BL, 9BC, 9BR, 9TL, 86, 133T, 239T; Corné Rautenbach 85T; Corrie du Toit 247; Dave Smallshire 72B; David F. Smith 37T, 42L, 48, 56, 103; David Hastings 125T; David Marvin 77B; David Monroy R 219; David Rees 122B; Dennis Farrell 184, 241T; Dennis Paulson 38B, 57T, 102B, 113, 119T, 236T; Didier Descouens/Muséum de Toulouse 243B; Dinesh Valke 45T; Eduardo Axel Recillas Bautista/ facebook.com/yolcatzin, @yolcatzin.mx 145B, 213; Erland Refling Nielsen 4, 5, 11, 21T, 25, 32-3, 35, 40, 41, 43T, 54T, 59T, 71TR, 75T, 78, 79L, 81, 83B, 88, 89, 92, 99T, 101T, 105T, 105B, 106, 107, 108, 110, 114T, 118, 119B, 122T, 123, 124, 137B, 139T, 139B, 141, 142T, 143, 152T, 153, 154B, 155, 158T, 163T, 163B, 167T, 178, 179, 187T, 191, 192B, 196, 197, 204, 222, 230, 231T; Fangshuohu 21C; gernotkunz 226; Giff Beaton 156; Graham Winterflood 82, 239B; Grete Pasch 218T; Han Onderwater 87B; Hanchongchong 208; Hans-Joachim Clausnitzer 146; Harald Schutz 72T; Héctor Ortega Salas 209; Ignacio A. Rodríguez 236B; istock/Marcophotos 142B; James A. Giroux 100; Jan van Leeuwen 223B, 228T, 229B, 247; Jean-Pierre Boudot 54B; © 2014 Jee & Rani Nature Photography 147T; Jens Kipping 14T, 99B, 151B, 169B; Jim T. Johnson 37B, 195, 223T; Jimmy Dee 93T, 97T; Jiri Hodecek 33; John C. Abbott/Abbott Nature Photography 29, 140; John Rosford 47T; Judy Gallagher 164B; Keith D.P. Wilson 63T, 66T, 90, 96, 128, 203; Klaas-Douwe B. Dijkstra 10, 20, 30R, 31, 42R, 49, 50B, 53, 95B, 217; Laitche 136T; Luiz Fernando Matos 145T; Marcel Silvius 188; Marcus F.C. Ng 28T, 28C, 28B, 34, 50T, 70B, 73, 101B, 125B, 147B, 150, 160, 161, 176T, 176B, 177, 185T, 185B, 190, 200, 201B, 202, 205, 206T, 211, 235T; Mark Sanders (EcoSmart Ecology) 79R, 109T, 207, 244; Martin Waldhauser 172B; Matt Weldon 111; Mike Melton 212; Mike Ostrowski 69B; Mildeep 46; Milind Bhakare 170; Natalia von Ellenrieder 228.
Nature Picture Library/ Andy Sands 240, Bernard Castelein 162, Konrad Wothe 181T, Jan Hamrsky 8, 167B, John Abbott 52B, 114B, 121T, John Waters 39, Jussi Murtosaari 151T, Lorraine Bennery 166, Michael & Patricia Fogden 157B, Nature Production 131, Nick Upton 32L, Oscar Dewhurst 22, Phil Savoie 62T, Ripan Biswas 19, Robert Thompson 12L, Ross Hoddinott 18B, Ross Hoddinott/2020Vision 135, Scotland: The Big Picture 21B, Sebastian Kennerknecht 157T, Steve Nicholls 18T, Sven Zacek 24, Thomas Marent 26-7, Willi Rolfes/BIA 134T.
Netta Smith 59B, 104, 227, 241B; Nick Moore 125C; Nicolas Mézière 12R, 13, 93B, 109B, 199C; Nuwan Chathuranga 234; Oleg Kosterin 67T; Onno Wildschut 120; Owen Strickland 127B; Pablo Martinez-Darve Sanz 94, 187B, 218B, 224R; Paul Cools 57B; Pedro Genaro Rodriguez 225B; Phil Benstead 192T, 233; Phil Chaon 115; Pierre Deviche 144; Prasan Shrestha 52T; Rasamoto 134B; Reiner Richter 138, 149B, 249; Renjusplace 136B; Rich Hoyer 194; Richard Orr 14B; Rick Buesink 186T; Rison Thumboor 36, 169T; Rob Felix 30L, 44; Robert Ketelaar 98; Rodrigo Conte 117; Rogério Ferreira 55B; Ruben Foquet 158B; Sayak_Dolai/India Biodiversity Portal 69T.
Shutterstock/ Danita Delimont 77T, Dave Montreuil 133B, GlashutteWinter 71B, JamesHou 97B, Nuwat Phansuwan 235B, Paul Reeves Photography 68, Ralfa Padantya 193, Zaksky 65T.
Stephen Richards 174; Stijn De Win 175; Summerdrought 198; TekuraDF 126, 182; Tim Faasen 221B; Tim Termaat 23; Tom Kompier 91, 95; Vengolis 62B; Victor Mozqueda 220; Vipin Baliga 171B; Vivek Chandran A 43B, 85B; Warwick Tarboton 245B; Wat Wongpan 121B; Wen Yuchuan 58, 130, 180, 181B, 199B, 201T, 214, 215, 216, 232, 237B, 245T; William A. Haber 248.

All reasonable efforts have been made to trace copyright holders and to obtain their permission for the use of copyright material. The publisher apologises for any errors or omissions and will gratefully incorporate any corrections in future reprints if notified.

ACKNOWLEDGMENTS

Threski Poorter, Dennis Paulson, and the team at Bright Press supported the book throughout. Its contents build on decades of knowledge, photos, and support provided by dragonfly enthusiasts such as John Abbott, Matjaž Bedjanič, Phil Benstead, Cornelio Bota-Sierra, Christophe Brochard, Sebastian Büsse, Seth Bybee, Viola Clausnitzer, Callan Cohen, Adolfo Cordero-Rivera, Rory Dow, Natalia von Ellenrieder, Sónia Ferreira, Günther Fleck, Heinrich Fliedner, Rosser Garrison, Benoît Guillon, André Günther, Matti Hämäläinen, Shantanu Joshi, Vincent Kalkman, Jens Kipping, Tom Kompier, Oleg Kosterin, Peter Maaskant, Milen Marinov, Andreas Martens, Pablo Martinez-Darve Sanz, Marcus Ng, Erland Nielsen, Bert Orr, Ângelo Pinto, Graham Reels, Michael Samways, Dattaprasad Sawant, Martin Schorr, David Smith, Netta Smith, Frank Suhling, Günther Theischinger, Jan van Tol, Reagan Villanueva, Jessica Ware, Marcel Wasscher, Beatriz Willink, Keith Wilson, Xin Yu, Wen Yuchuan, and many, many others. Please keep up the good work!